Protein

Protein

THE MAKING OF A
NUTRITIONAL SUPERSTAR

———

Samantha King and
Gavin Weedon

DUKE UNIVERSITY PRESS

Durham and London   2026

Project Editor: Ihsan Taylor
Designed by Courtney Leigh Richardson
Typeset in Freight and Retail by Westchester Publishing Services

Library of Congress Cataloging-in-Publication Data
Names: King, Samantha, [date] author | Weedon, Gavin, [date] author
Title: Protein : the making of a nutritional superstar / Samantha King and
Gavin Weedon.
Description: Durham : Duke University Press, 2026. | Includes
bibliographical references and index.
Identifiers: LCCN 2025029824 (print)
LCCN 2025029825 (ebook)
ISBN 9781478032922 (paperback)
ISBN 9781478029489 (hardcover)
ISBN 9781478061694 (ebook)
ISBN 9781478094654 (ebook other)
Subjects: LCSH: Proteins | Proteins in human nutrition | Nutrition
Classification: LCC QP551 .K4925 2026 (print) | LCC QP551 (ebook) |
DDC 572/.6—dc23/eng/20250929
LC record available at https://lccn.loc.gov/2025029824
LC ebook record available at https://lccn.loc.gov/2025029825

Cover art: Courtesy Shutterstock/Boule.

publication supported by a grant from
The Community Foundation for Greater New Haven
as part of the URBAN HAVEN PROJECT

Contents

Acknowledgments

It's been almost a decade since we started to question the outsize role that protein has come to occupy in contemporary food and fitness cultures. As the planet has lurched from one crisis to another, protein has been one of the few constants in our lives, as have the precious relationships and communities that have supported our work along the way.

Since we live on different continents, we have spent countless video calls ruminating on protein matters, a practice we began long before the pandemic propelled academics en masse into the remote work environment. In this context, conferences and workshops became especially important venues for developing and testing our ideas. We are grateful to Mary McDonald and Jennifer Sterling for hosting us in their Sport, Technology, and Society sessions at the annual conference of the North American Society for the Sociology of Sport (NASSS), starting in 2015 and through several subsequent iterations; each meeting pushed us to take our analysis one step further. Dave Andrews's invitation to a workshop at the University of Maryland provided an occasion to garner early feedback and the provocation to figure out what happens to protein once we ingest it. Participants at the New Materialisms and Environmental Humanities Conference in Paris helped us clarify our theoretical trajectory, as did those at the Qualitative Research in Sport, Exercise and Health Conference in Vancouver, and the European Society for the Sociology of Sport Conference in Bø. The audience at the Alan Ingham Memorial Lecture at NASSS in Montreal allowed us to speak more directly to the scholarly community with whom we have the most enduring connections.

A special thank-you goes to our generous book workshop participants, Mary McDonald, Josée Johnston, and Madeleine Pape, whose astute and detailed comments greatly enhanced our analysis. Mary Louise Adams, Chris McGlory, Radhika Mongia, Ishita Pande, Paige Patchin, two anonymous reviewers, and members of the Duke University Press board provided thoughtful and challenging feedback. These interlocutors' careful readings of our manuscript have sharpened it significantly.

Many colleagues, friends, and family members who may not have read the work have spent time discussing all things protein with us or otherwise supporting our work. Great thanks are due to Adam Ali, Denita Arthurs, Chris Bongie, Ali Bowes, Laura Bradley, Stuart Brown, Sheena Clowater, Martin Danyluk, Judy Davidson, Pam Havery, Michelle Helstein, Jake Hodder, Kathy Jamieson, Susan Lord, E. MacDonald, Kat Mazurok, Brad Millington, Michele Murray, Ishaan Pande, Kip Pegley, Elaine Power, Carolyn Prouse, Vicki Schmolka, Courtney Szto, Anna van der Meulen, Matt Ventresca, Patricia Vertinsky, Devra Waldman, Sandy Wells, Bobby Whyte, Brian Wilson, Liv Yoon, the Queen's Coalition Against Austerity community, and the King, McClenaghan, Patchin, and Weedon families.

Illustrator Shawn Forde and image curator Anouk Jonker patiently processed our sometimes garbled artistic visions into wonderfully rich and evocative images. Research assistants Izzy Altoé, Jessica Guilbault, Gagan Minhas, Christine Moon, Amanda Morales, Natalia Mukhina, Gözde Öncil, Andrea Reid, Kaitlyn Seow, Sarah Smith, Jill Takacs, and Grace Wedlake provided invaluable contributions as we strove to keep the project moving forward.

We are grateful to Courtney Berger at Duke University Press for her belief in the project and her encouragement and enthusiasm throughout the process. A big thank-you to Laura Jaramillo, Erika Jackson, Christi Stanforth, and Ihsan Taylor for their keen editorial support. It has been a pleasure to work with the team at Duke.

Funding from the Social Sciences and Humanities Research Council of Canada supported the project between 2018 and 2023, first with an Explore grant and later with an Insight grant. Nottingham Trent University's Department of Sport Science underwrote a research visit to Canada with Quality-Related funds that facilitated a fruitful spell of thinking and writing together. Queen's Scholarly Publishing Fund and a generous head's research stipend are helping to ensure the book gets read by as broad an audience as possible.

To Willy Vestering and Paige Patchin for your unceasing support, patience, and curiosity, we cannot thank you enough. If anyone can testify to protein's persistence, you can. Last but not least, we are grateful to Sacha Andreas and Sprout, who became part of our worlds in blissful ignorance of this book.

INTRODUCTION: PROTEIN

A Solution in Search of a Problem

In the contemporary nutritional imagination, protein assumes center stage. Touted as a muscle builder, a weight-loss aid, an antiaging fix, and more, this dietary superstar is the subject of an inescapable cultural preoccupation, even for the most hardened of skeptics. We count ourselves among the latter. As critical scholars of health and the body, questioning disciplinary edicts about what to eat is our bread and butter. But when we noticed that we, too, were getting caught up in the "protein talk" that pervaded the fitness cultures to which we belong, we began to think more carefully about the origins of this particular obsession and what it represented. While we were curious about the longer history of protein's preeminence, our initial questions focused on its current status: What is the appeal of those giant plastic tubs of highly processed protein powder for our well-fed, "clean eating" gym buddies? Why do food manufacturers emphasize the protein content of their products in increasingly larger fonts? What keeps consumers rotating through the latest high-protein dietary regimen, from Atkins to paleo to Dukan? Who needs protein added to their beer, potato chips, or ice cream? Why are people scared of carbs?

Among our most pressing questions was how to situate the fixation on protein in relation to widespread concern about the role of industrial animal agriculture in the climate crisis.[1] As global meat and dairy production continues to rise, its impact on greenhouse gas emissions, deforestation, pollution, biodiversity, animal welfare, and human health has prompted growing calls for a "protein transition" away from resource-intensive farming. The popularity of plant-based diets in traditionally meat-centric locales—including the United States and Canada, where our focus lies—is indicative of this push. But if anything, the specter of protein deficiency looms even larger in vegan discourses, or at least in the marketing of industrial plant-based foods, where the promise of high-protein content is requisite, perhaps offered as reassurance to consumers unable to imagine adequate sustenance in the absence of animal flesh. The all-caps guarantee of "20G OF PLANT PROTEIN PER SERVING" that appears in every instance of promotional material for the Beyond Meat burger exemplifies protein's allure across dietary regimes and ethical registers. In the alt-meat world, it is less high-protein diets that garner scrutiny and more the specific ingredients they contain. This orientation aligns with the growing interest of large food conglomerates in alternative protein production and their efforts to rebrand themselves as suppliers of "sustainable protein" rather than makers of meat.[2]

Gifted with a halo effect brighter and more enduring than any other category of nutrient, protein can do no wrong. The climate crisis might be motivating efforts to rethink meat, but not the foremost place of protein in our diets. Unlike with carbs or fats, there is little room for the notion of "bad" proteins in nutritional discourse. While low-fat and low-carb diets are ubiquitous, low-protein diets are virtually unheard of outside the specialized medical realms in which some patients with impaired liver or kidney function are treated. With its "near-universal" attraction, protein has become a multipurpose fix.[3] As one food marketing director puts it, "Protein's continued growth is driven by the fact that it appeals to all demographic groups. . . . Fitness for Millennials and Gen Xers, energy and weight management for boomers, wasting and muscle loss prevention for seniors."[4] The sociological subtexts of this message? In cultures of overwork, exhaustion, and anxiety, protein is understood to fuel vigor and vitality, despite the scientific consensus that carbohydrates are the body's primary sources of energy. And in antifat and ageist cultures obsessed with body size, shape, and composition, protein is a win-win prospect, especially for corporations that can sell the same protein-fortified products to groups of consumers with opposing goals: Want to lose weight? Eat more protein! Want to bulk up? Do exactly the same thing![5]

Going into the project, we imagined that these seemingly magical properties might relate to the conventional idea, rooted in biochemistry, that proteins

FIGURE I.1. A multipurpose fix? Protein at the grocery store. Illustration created by Shawn Forde.

are the "building blocks of life" and that proteinous foods are essential for human bodies to grow and repair. We weren't wrong—protein's known and hypothesized biochemical capacities are crucial to its contemporary cultural allure and economic value. Yet we have also come to realize that after almost two centuries of research, there is little scientific consensus about exactly how much protein people should eat for optimal health and whether those who consume adequate calories should pay any special attention to their protein intake.[6] Even the seemingly indisputable claim that protein constitutes a crucial component of the human diet gives us pause, given all it assumes about what this substance is and what it can do.

Toward some stable ground, it is more accurate to say the following: All animals, including humans, need nitrogen to make the proteins that perform multiple bodily functions. Nitrogen is important to underscore here. When used to refer to substances that humans ingest, "protein" is a proxy for the nitrogen content of food—nitrogen that is then used to make proteins once it enters the body. At a molecular level, amino acids are the primary vehicles through which we obtain nitrogen. This is why "protein" is also a proxy for amino acids, though amino acids are additionally composed of hydrogen, carbon, and oxygen. The number of nitrogen-containing, protein-building amino acids the body requires is usually identified as twenty. Some researchers, however, put that figure at twenty-one, to account for an additional "nonstandard" amino acid, selenocysteine, discovered in the late twentieth century, or at twenty-two if they are including pyrrolysine, another rare amino acid identified in 2002. Of those proteinogenic amino acids, nine are categorized as indispensable, which means that they cannot be produced by the body and must be obtained exogenously, from food.

This is the point at which the possibility of making overarching claims about protein intake begins to fray. In the face of inconsistent, imprecise, and unreliable experimental techniques for measuring either protein metabolism or the protein content of particular foodstuffs, questions about the form in which people should obtain those indispensable acids, in what quantity, in relation to which health conditions or aspirations, in the context of which food systems, and at which point in their lifespan (or indeed their day), remain the subject of ongoing research and debate.[7] This contentiousness is not readily apparent outside the nutrition research world or fitness and diet circles, where discussions about the desired level of protein intake for optimal performance and weight loss are de rigueur, and strong and often divergent opinions abound. If you are not privy to such conversations and turn to the internet for information, your findings will depend on where you are located. If you're searching from Canada or the United States and you ask, "How much protein should I eat per day," your results will turn up the same piece of apparently uncomplicated advice repeated over pages of results: "0.8 grams of protein per kilogram of body weight." But if you are in the United Kingdom, where food consumption and availability patterns are similar to North America, you will be advised to consume 0.75 grams per kilogram. If you are in Japan, you will be advised to consume 1.08 grams per kilogram, and if you are in one of the many countries that use either World Health Organization or European Union guidelines, the amount sits at 0.83 grams per kilogram. These differences are small but significant, reflecting not just contextual variables related to food availability and culinary cultures but also epistemological uncertainties and political contestations.

In the United States and Canada, the recommended daily allowance (RDA) of 0.8 grams of protein per kilogram of body weight is derived from the Dietary Reference Intakes (DRIS), a set of nutrient reference values released in 2002 by an expert panel appointed by the US National Academy of Medicine.[8] Used by policymakers, healthcare providers, and the food industry to plan and assess diets for healthy individuals and populations, the DRIS are powerful tools for the dissemination of nutritional norms and the determination of what people eat. They underpin the *Dietary Guidelines for Americans* and *Canada's Food Guide*, they form the basis for food assistance and school lunch programs, they constitute the standard against which the world's 250 million subscribers to *MyFitnessPal* can track their daily nutrient intake, and they are used by food and supplement manufacturers seeking to market their products. But if one actually ventures to open the document, any sense that the DRIS for protein are straightforward quickly evaporates. Spanning 179 pages, the "Protein and Amino Acids" section of the DRI report is highly elaborate and endlessly complex.[9] It is evident that the DRIS emerge from reproducible empirical data on the protein needs of humans—they are not without foundation, in other words, or contested to the point of meaninglessness. But the report is honest about the limits of contemporary nutritional knowledge, noting at the outset that "proteins in both the diet and body are more complex and variable than the other energy sources, carbohydrates and fats."[10] Thus, even in the realm of official dietary guidelines, where specifications are set by expert committees after years of deliberation, recommendations about protein consumption are tentative.[11] They are also subject to ongoing pressure from interested parties such as agribusiness, industry-sponsored researchers, and health evangelists of various stripes who would like to see them shift in one direction or another.[12]

Beyond the consensus statements that underpin dietary guidelines, what agreement there is about protein intake converges around the fact that protein deficiency is extremely rare in the absence of severe hunger and almost nonexistent among the consuming classes at the center of the protein boom.[13] Garth Davis, a bariatric surgeon and coauthor of the diet exposé *Proteinaholic*, conveys this point with a scenario featuring a five-foot-ten man weighing 160 pounds.[14] If all this man ate at each of his three daily meals was thirty Lay's potato chips and two slices of Domino's plain pizza, Davis contends, he would be consuming 1.6 more grams of protein than his daily recommended allowance. While the precise level at which daily protein intake should be fixed remains unsettled, the larger point stands: Protein is in practically everything people consume, from lettuce to potatoes, and if they have enough food to eat, they must go to extreme lengths to become protein deficient. It thus strikes us as more than curious that those people who appear to be most fixated on their protein intake are those who have the least to be concerned about. How and why, then, has protein emerged as an über-nutrient in the satiated world? To what ends? And at what costs?

Situating the Protein Boom

One way to understand contemporary protein mania is in relation to the broader rise of what critics call "nutritionism."[15] This is the ideological process through which the value of food is reduced to its biochemical components and measured according to Eurocentric, scientized standards rather than taste and experience. Growing out of a complex assemblage of forces, including the profit imperatives of global agribusiness and the responsibilizing imperatives of state-sponsored dietary guidelines, the hegemony of nutritionist thinking is undoubtedly at work in the extraction and isolation of protein from its nutritional and social context and the reduction of its value to that which can be assessed and measured as the central part of any meal.[16] Nutritionist logic thus oversimplifies the role of individual nutrients both in the pursuit of bodily health and in the solutions to the systemic problems (e.g., social inequality) in which nutrients are or appear to be implicated—a recurring issue in the story of protein, as we shall see.

Thinking about the broader social and historical conditions in which nutritionism has taken hold leads to another—complementary—way we might explain protein's allure, which is through the lens of what sociologist Nikolas Rose calls the age of "molecular biopolitics."[17] Rose argues that during the mid-twentieth century, scientific understandings of life and the biological body were "molecularized," which is to say that the body was reconceptualized at submicroscopic, subcellular levels and thus became knowable at a completely different scale than in previous eras of scientific investigation. Compare, for example, long-standing knowledge about muscles and their anatomical function in moving and powering the body to contemporary knowledge about amino acid chains and their genetic functions in the creation of muscle tissue, a process that is discernible only at the molecular level. Rose's claim is not simply about the scientific and technological developments that make this knowledge possible but also about their economic and social reach, not least the circuits of capital that are allied to new modes of knowing bodies beneath the skin.[18] His argument suggests that in the molecular age, optimizing our genetic, hormonal, and neurological selves in ways that mitigate individual risk and maximize the potential for flourishing and longevity has become possible, at least for those with access to this knowledge and the means to act on it. Health in these formulations is not just a matter of population governance or disease prevention but of personal actualization at the molecular scale, often cast in the language of the molecular sciences. Consider how particular amino acids are now paired to particular health, fitness, and aesthetic goals, enabling consumers to pursue precise, targeted, "optimal" nutrition—L-serine for brain health, leucine for muscle mass, collagen peptides for hair, skin, and nails, and so on. The sheer possibility of referring to a nutrient as a superstar, granting it a cultural appeal, and incorporating it into the ways people understand their own bodies and relationships to food, owes a great deal to these shifts in scientific knowledge and the senses of selfhood they have enabled.

These are frameworks to which we return in the analysis that follows, but neither nutritionism and the emergence of the "nutricentric subject," nor molecular biopolitics and the emergence of the "neurochemical self," shed light on the specificity of protein primacy in contemporary dietary culture.[19] Why is protein, rather than fat, carbohydrate, vitamin D, or any number of other nutritional categories, so fetishized in the present moment? Why is *this* molecular matter the resounding staple of dietary regimens that have varying, and at times competing, objectives?

In response to these questions, we propose an additional context that helps explain the protein boom, one which is likely familiar to those already curious about or invested in protein's ubiquity. Since the 1980s, numerous studies have explored what historian Jürgen Martschukat calls the "age of fitness" in Western societies.[20] The popularity of practices such as running, yoga, and weightlifting, and the growth of associated industries, is the most obvious expression of the explosion of interest in the pursuit of a fit, healthy body that marks the last half century.[21] As for what animates this investment, scholars of physical activity have identified a litany of factors, from raced, classed, and gendered expectations of body shape and size, to antifatness stoked by obesity research and policy, to the sedentary nature of postindustrial societies, to a growing evidence base for the psychosocial benefits of physical activity.[22] Most critically attuned researchers home in on neoliberalism, however, as the overarching political and economic condition of possibility for the emergence of the fitness boom.[23] The affinities between fitness and neoliberalism are certainly compelling, given the latter's emphasis on individual responsibility for health and the former's role in helping to instantiate and legitimize the commonsense notion that individuals are in control of their corporeal destinies.[24] Fitness as a bodily practice and neoliberalism as a mode of governance also share a declared aversion to unproductive excess—be it fat or fiscal "inefficiencies" associated with redistributive state spending that benefits collective well-being. Both also ascribe a moral quality to entrepreneurial activity that bears evidence of discipline, effort, and self-control. In this confluence, neoliberalism begets and explains the valorization of fitness in contemporary societies and, in particular, an aesthetics of bodily productivity that protein supplementation is believed to facilitate. If we accept the "age of fitness" as the embodied cultural expression of neoliberalism, then we might posit protein, with its flexible capacities and transformative promises, as an ideal vehicle for nourishing fitness regimes. It is not difficult to see how protein-enriched bodies forged through physical activity can connote value and (self-)worth in cultures saturated with neoliberal values, especially when allied to the prevalence of discourses of nutritionism and molecular selfhood.

The reach of neoliberal norms is such that fitness imperatives affect all of us, including those who don't work out. As physical activity consciousness, practices, spaces, knowledges, technologies, fashions, media, and diets infiltrate more and more of daily existence, weighing even on those who do not engage in instrumentalist exercise, so too does the multibillion-dollar market for protein supplements and fortified foods. Indeed, if the growth of fitness now amounts to what we might conceptualize as the athleticization of culture, then we wager that the athleticization of food culture, and especially the proteinization of food culture, have been major ingredients in this shift. It is not just that once-niche foodstuffs ingested by serious bodybuilders have entered the mainstream fitness world—a point we elaborate in chapter 5. Rather, specialized and highly technical approaches to diet that were once followed only by elite athletes seeking fractional improvements in performance have spread far beyond the world of fitness, thus revealing fitness as a primary vehicle for the spread of nutritionism, molecularization, and neoliberalism—all key factors in protein's rise.

These insights notwithstanding, it is worth emphasizing again—and not for the last time—that protein and its popularity resist and exceed attempts at neat and tidy explanations. In this instance, temporality emerges as an issue, since the protein boom is not distinctive to neoliberalism, and the present era is not the first to witness a fitness boom.[25] While protein's contemporary status is unprecedented in market terms—the protein powder market alone is estimated at US$21.6 billion[26]—protein has been enlisted in economic and political projects since its establishment as a nutritional category in mid-nineteenth-century Europe, and in numerous agro-technological, financial, and biopolitical ventures in the ensuing two centuries. The point here is that the obsession with protein has roots in global expansionist processes and projects that complicate any attribution to a single social formation or historical conjuncture. The last two chapters of the book explore the present political moment as indicative of this historical contingency and complexity: As neoliberal orthodoxy flounders and protofascist populist movements gain strength all over the world, including in the United States, where a new age of "muscular capitalism" has taken hold, many online outlets for vindictive grievance politics and sovereign self-making projects are fueled by the supplement industry—a context we label "the muscular manosphere."[27] What is extraordinary about protein is not just that it plays an outsize role in powering this noxious brand of muscular capitalism in the digital networks in which it has flourished, but also that it has assumed this role before in quite different geopolitical and sociocultural contexts, as we chronicle in the chapters to follow. What kind of substance can withstand this turbulence, let alone thrive in it?

The need to think carefully about protein's remarkable staying power was underscored to us when the story of the protein boom entered a new chapter in the twenty-first century, one marked by an explosion of interest in plant-based and cellular meats. In response to escalating concern about the climate emergency, unstable supply chains, waning food security, and human and animal well-being, meat analogues are being pursued and touted as healthy, humane, and sustainable alternatives to large-scale livestock farming.[28] Scholars from many disciplinary vantages are asking critical questions about these developments and the multitude of possibilities and problems they raise.[29] Their many lines of inquiry relevant to our interests include the economic, political, and social forces driving—and obstructing—a protein transition, and the diffuse and uneven developmental pathways of alternative protein technologies, which range from niche innovations in cellular agriculture to the refinement of long-standing legume- and wheat-based alternatives produced by conventional food and commodity processing conglomerates as well as smaller corporate actors.[30] Other relevant studies explore the increasing concentration of power in capitalist agribusiness under the banner of "protein";[31] the construction of metaphors and "promissory narratives and future imaginaries" about alternative protein technologies among the biotech start-ups leading the development of cellular agriculture;[32] the importance of patentability and the reframing of "food as software" in the high-tech alternative protein sector;[33] the contested discursive terrain around the safety, feasibility, and sustainability of novel foods;[34] the ontological status of fake meat and dairy and their entanglements with animal agriculture;[35] the moral, affective, and aesthetic considerations raised by cellular agriculture;[36] and the environmental and health consequences of a shift to novel protein production and consumption.[37]

As we were finishing the book, new research on shifting expectations about the potential of alternative protein development to challenge food system orthodoxies was appearing.[38] Public prospects have been tempered somewhat since 2022, when the market for the new generation of plant-based meats such as the Beyond Burger and the Impossible Sausage turned out to be more constrained than their promoters and investors initially hoped.[39] The collapse of their share price, and their removal from the menus of some of the big fast-food chains was followed, in 2023, by the closure of New Age Eats, a cultivated meat start-up that shuttered its doors having failed to attract the necessary funding to support ongoing infrastructure development.[40] Academics and journalists who are tracking the trajectories of alternative proteins point to a range of reasons for these challenges and flops. They usually land on some combination of fiscal, logistical, sensorial, and health drivers not helped by a meat industry–backed campaign that depicted burger and sausage analogues as packed with hidden additives and therefore less healthy than beef or pork—a move to which we return in the epilogue. Critic Julie Guthman is particularly uncompromising in her assessment of what she depicts as ill-informed and ineffective tech solutions that fail to address the root causes of a broken food system.[41] Silicon Valley is Guthman's focus, though this is only one locus of development in alternative proteins, and recent upheaval in the United States constitutes only one chapter in a story that continues to unfold along multiple pathways, including in the United Kingdom and Germany, where governments are starting to invest in a sector that has until recently been funded largely by private money.[42] At the same time, some protein start-ups and innovators have thrown their lot in with the same corporations charged with promulgating the harms of industrial animal agriculture, given the presumed capacity of these firms to scale up production and infrastructure.[43]

From our vantage point, what is striking about all this activity is that protein's rise as the dietary elixir par excellence and its status as a nutritional superstar remain largely unquestioned, even by those seeking systemic transformation.[44] As entrepreneurs and scientists attempt to engineer a future in which food is cleansed of its associations with the harms of industrial animal agriculture, and critical scholarship appraises their claims, the anchoring category of their engagement gets naturalized and secured. Whereas anxiety about the sustainability of the food system and efforts to transform it have forced a deep reckoning with the meanings and materialities of meat and dairy, the same cannot be said of protein, which emerges as an escape valve of sorts for a range of actors seeking to rethink and rebrand animal sustenance. In these conversations, protein tends to hover hazily in the background, its capacity to attract an ongoing supply of attention and investment, on the one hand, belying its elusive and abstract character, on the other.

Materialisms Old and New: Theorizing Protein's Capacities

To grasp the significance of protein's hold on the nutritional imagination requires figuring its contemporary status as more than the conceit of empire-building nutritional science, more than another example of nutritionist reductivism or molecular embodiment, and more than the sculpted silhouette of neoliberal capitalist hegemony reflected against the gymnasium wall. The manufacture of cellular and plant-based proteins orients attention to the *active* material capacities of the substances gathered under this category, the vital force with which protein is synonymous, alongside the problems of the global food system and the ecological breakdown to which entrepreneurs are responding. The interest in novel proteins also draws attention to the multiplicity of meanings adhered to protein and the benefits it is intended to generate, as well as the history of protein's uptake over the *longue durée* of its biography. What's required is an approach that brings together protein's socio-ecological history, constitution, and potential, one that attends to its material manifestations in order to explain its historical emergence as a coherent and lauded category.

Our approach and arguments in this book come through engagements with a host of scholarly fields that take seriously relations among power, nature, and history. Political ecology, science studies, food studies, sport studies, and body studies have been especially key in lending insights about how the vital capacities of biochemicals are connected to environmental concerns, how metabolic processes are implicated in food politics and vice versa, how knowledge about diet is constructed and interpreted, and how economic and social concerns cohere in embodied practices. Two schools of materialist philosophy that respectively emphasize the force of world-historical processes in shaping social relations, and the force of material life to act and make a difference in those relations, have been especially important. To unravel their explanatory power and their theoretical continuities and tensions, let us foreground the form in which protein's allure first made itself known to us.

One of protein's primary manifestations in the contemporary world is as a nutritional supplement, an object for consumption in health and fitness cultures. When we call protein an elixir, we conjure its quasi-magical associations with good health, even as we argue that protein is much more than any single ascription can hold. Indeed, one of its tendencies is to exceed categorization and veer outside the crosshairs of analysis. But we have to start somewhere, and protein as a valuable commodity in the nutrition supplement market is the form with which many readers of this book will already be acquainted and the form that primarily occupies us in the chapters that follow.

Studying protein through a materialist lens means recognizing that when a gym-goer, dietary enthusiast, or unsuspecting consumer ingests proteinous foodstuff, that abstracted matter *comes from somewhere*, and the transitory moment of consumption connects them to a dense web of lives and forces beyond their actions and milieus. This is the same logic that Karl Marx deployed in his decision to lead with the proliferation of commodities in *Capital*, and although his work derives from another century, it is a mode of materialist thought that continues to inform critical accounts of social and cultural life.[45] Marx was writing through the upheavals of industrializing societies in mid-nineteenth-century Europe, where the scale and type of production was transforming landscapes, livelihoods, and labor all around him. Observing capitalism's alchemic transformations of earthly matter into commodities, Marx set about tracking those transformations into discernible social relations so as to demystify the workings of capital. There are many specific philosophical and economic debates to be had about this method, but for us it is the general approach to situating commodities that has proved most instructive. For example, in chapter 1 this mode of analysis helps us discern how protein's invention as a singular nutrient by a vanguard of nineteenth-century European biochemists was, from the outset, a process of commodification, one tethered to the expansionist imperatives of capitalist and colonial world-making that has now reached an apotheosis in the circulation of protein in global food markets. That chapter tells the story of how renowned biochemist Justus von Liebig—whose research on soil was a significant influence on Marx's own thought[46]—held together the ontological question about what exactly protein is with the entrepreneurial matter of its deployment as a necessary supplement in domestic and international markets. Here, we work in a similar way, reckoning with protein's biochemical and commodity forms in order to situate them within circuits of exchange and accumulation, chart the social conditions of their emergence and diffusions, and demystify the seemingly endless and often unmitigated stream of "protein hype."

This "older" materialist approach for following commodities and situating their circulation in world-historical processes imprints on the entirety of our analysis, as does the broader sensibility toward illuminating how systemic forces have animated protein in the pursuit of power and profit. They aid us in "following" our subject on its travels through laboratories and public health campaigns, sewers and pharmaceuticals, gyms and digital networks, and the bodies of flesh, land, and water that protein, in turn, reshapes along the way. In this sense, our work resonates with kindred literature on the pursuit of "things" through different spaces, divisions of labor, social relations, more-than-human actor entanglements, and other material-discursive webs.[47] Arjun Appadurai's insight that "all things are congealed moments in a longer social trajectory" has proven especially useful when infused with a method that opens up the social to include the ecological, multispecies trajectories along which protein travels.[48]

No doubt owing something to these analytical tools, our early attempts to follow protein and to pin down its composition and impacts engendered several scoops of skepticism about its transformative properties. Yet the more we learned of protein's indeterminacy and evasiveness, even in the face of commodification, the more we recognized the need to take these qualities seriously. After all, proteins are many things, from molecules to meal staples and much more besides. They are in us and around us; they move between us, act on us, and operate beyond us. Protein can take the form of viral "spikes" that penetrate cells with dangerous consequences for human health or of granular powder mundanely consumed after a morning run. Any attempt to make scientific, health, or commercial claims about protein thus requires a great deal of boundary work just to delimit what "it" is in that instance, let alone its effects in any given context.

As we followed protein on its circuitous travels through markets, cultures, and bodies, the significance of its dexterity and transformative capacity gradually began to emerge less as a paradox than as a potential explanation. Might it be that this elusive, indeterminate quality is no mere glitch in knowledge about protein but could actually be central to its power? This proposition about protein's capacities aligns with what is often described as a "new" mode of materialist thought—a distinct collection of philosophies and research practices that have become influential in the last two decades or so. New materialist approaches (alongside related approaches such as posthumanism, actor-network theory, and speculative and agential realism) seek to theorize the agency and sociality not just of humans but of plants, animals, chemicals, machines, and other entities whose vitality and affective capacities, it is often implied, have hitherto been neglected.[49] In some cases, such recognition is mobilized to theorize the human condition at a moment of profound and ruinous ecological transformation known widely as the Anthropocene.[50] Recognizing the agentic capacities of material life serves as a corrective to the idea of both agency as a primarily human capacity and the Anthropocene as an epoch in which human activity is equated to the "great forces of nature" in its reach and effects.[51] This may constitute the unifying political intervention of the "new" materialisms. Their influence is seen in an outpouring of concepts that extend the boundaries of human subjecthood, afford agency to nonhuman phenomena, and aim to attune readers to a shared planetary existence of multispecies, multiorganismic relationality.

Debates continue to rage about both the newness of this version of materialism and the appropriateness of the Anthropocene as an epochal denominator of our socio-ecological present.[52] In sweeping terms, we might say that "old" materialist perspectives emphasize the force of history and hierarchical systems of power, whereas "new" materialist perspectives emphasize the agency of matter and nature in constituting social and ecological life. In practice, these approaches traverse such binary framings, but the extent to which scholars integrate these considerations varies greatly, and there has been significant discord between their more ardent adherents. This conflict reflects long-standing questions about the philosophical premises of Marx's historical materialism and its contemporary relevance, on the one hand, and a suspicion that the "new" materialisms fail to adequately reckon with the histories of race, capital, and empire that have shaped so much of the modern world, on the other.[53] Whereas the newer schools stress the distributed agency of forces involved in climatic upheaval, in part to temper fantasies of mastery over nature that still undergird some Anthropocene theorizing, their critics point out that such assessments often underplay the human, social dynamics of climate transformation, and the fact that climate change is neither waged nor experienced uniformly by all groups of people. Similarly, some hold that the Anthropocene demands fresher modes of understanding the world than are offered by the established conceptual repertoires of the interpretive social sciences and humanities.[54] Others, meanwhile, see as obvious the need to incorporate several centuries of knowledge into contemporary approaches and argue that the fetishization of the "new" risks neglecting the ongoing violence inflicted by capitalism and colonialism as the world heats up.[55]

Allied with traditions of thought that have never imagined human corporeality outside of nature, our approach in this book stems from a conviction that understanding the vital capacities of protein need not detract or distract from its implication in world-historical processes.[56] On the contrary, one of our central contentions is that protein's elusive, dynamic character is the key to its social power. Far from being incidental to its hallowed status and cultural prevalence, protein's vital capacities and opaqueness are what make it amenable to biopolitical, economic, and environmental imperatives. In alignment with scholars who conceptualize agentic matter not beyond but *as* power and history, this claim strikes a dialogue with "new" materialist and related approaches that explore how lively, multispecies, multiorganismic relationships constitute "human" life, while retaining an emphasis on the social forces that shape and mediate protein biologies, with far-reaching and disparate effects.[57] From this vantage point, we explore how protein has been put forward as the magical solution to a diverse range of problems: from malnutrition in the Global South in the form of culturally alien techno-foods, to pollution in industrial farming in the form of regenerated whey waste, to muscle loss in seniors in the form of drugs and other prescriptions designed to increase protein synthesis, to the "crisis" of contemporary white manhood in the form of bodybuilding supplements, and to the climate emergency itself in the form of "green" or "alternative" protein. We situate each of these innovations in their historical and socio-ecological contexts, highlighting colonial superiority, capitalist ingenuity, age-as-deficit, white Trumpian masculinity, and ecomodernism as key to understanding protein's attractiveness as a multipurpose fix for all manner of ills, real and imagined.

The need to think of protein not only as a substance embroiled in hierarchical and often violent processes of commodification but also as an agentic, socio-ecological phenomenon did not arise from philosophy alone. It came through our exploration of the creation of whey powder, initially familiar to us and likely to many readers as the most popular protein supplement. In seeking to chart whey's commodification and appeal, we discovered that it has a curious double life, acting not only as a health elixir but also as a noxious by-product of the dairy industry.[58] The transformation of whey into a palatable, proteinous foodstuff has been triumphantly touted as a green solution to the intractable problem of toxic whey waste. While that waste is itself an outcome of dairy overproduction, it gets reframed in this story as an opportunity for the exercise of technoscientific ingenuity and capitalist innovation—its transformation into a profitable commodity demonstrating the possibility of making a market out of anything. On closer inspection, though, we found that the nitrogenous quality that allows whey to materialize as both panacea and poison does not disappear through mass production or even upon digestion. Rather than being purified through its commodification and then disappearing after ingestion, whey reemerges in other guises once it has passed through the bodies of the humans who consume it. In other words, whey's dense capacities present challenges to the global nitrogen cycle not just at the point of production, where whey powder is posited as a solution, but to the lands and waters it moves through after consumption and excretion.

Figuring protein as a socio-ecological phenomenon might appear paradoxical given that proteins are conventionally understood as fundamentally biological and that human biology is typically understood as unfolding largely "inside" the body. But the lessons we gleaned from pursuing these dynamic molecules on their travels through a plethora of bodies and substances dispelled any pretensions to this duality. We traced whey's movements from cows and their milk into supplements and additives, and on into humans and other animals who consume whey as part of their diet. Given that whey waste is only ever partially recaptured and purified, we were also led from dairy wastewater into lakes, rivers, and oceans, where it spurs plant growth and denies fish and other aquatic creatures the oxygen they need to survive. As a result, we were impelled to understand how whey exceeds the boundaries of the body and the environment, the ecological and the social, the healthy and the hazardous. The same, we later realized, was true of protein writ large: If we wanted to understand where the protein people consume comes from, where it goes once their bodies have processed it, and how its vital capacities make a difference in the world, we would need to consider protein in its diverse manifestations, from the cellular to the social.[59]

Protein emerges from this analysis not as a purely biological substance, a neutral nutritional category, or a singular cultural artifact but as a powerful, shapeshifting, socio-natural agent enlisted in diffuse webs of political, economic, and ecological forces. This is a conceptualization that denaturalizes the category of protein and destabilizes its axiomatic associations with health and nutritional value while accounting for the difference that protein's lively attributes, affective propensities, and relational qualities make in the world. It is an orientation that takes seriously protein's materiality, not as part of a philosophical exercise to determine its thingness, or a biochemical exercise to explain its role among the "building blocks of life," but in order to understand how its vital properties (i.e., what makes it both nutritionally valuable and environmentally problematic) are mobilized within iniquitous entanglements. It is a way of knowing protein as a substance that is easily adhered to diverse corporeal, economic, and political ends precisely because of its extraordinary dexterity, and it is an approach that keeps in view our subject's capacity for comprising, sustaining, and fortifying biological life while recognizing that little about protein is stable, self-evident, or neutral, whatever shape it takes.

In keeping with this emphasis on relationality and materiality, it bears emphasizing that protein's capacities are neither exercised autonomously nor simply willed into being by the might of protein hype or the powerful imperatives that propel protein markets. The active body, the labor it exerts, and the biophysical processes that follow are vital to materializing protein's capacities. This is something that all protein enthusiasts know, yet it is easily overlooked. In order to have a chance of realizing the promises with which it is endowed, in order to process its nitrogenous potential and transform it from food into muscle, protein requires sufficient physical and metabolic work. Put differently, protein's continued appeal rests on the specter of its absence and the desires, anxieties, and persistence of fitness enthusiasts and others subject to its allure. Without a preoccupation with muscle mass and strength, and with weight loss and gain, the labor required to unlock protein's capacities would occur only incidentally and certainly not in a way that would support the booming transnational market in protein products we see today. In other words, the hype surrounding the protein boom is hollow without multifarious forms of physical and affective labor and a populace willing and able to bring about those metabolic transformations. A socio-ecological approach to protein engenders a sensitivity to the biosocial labor of bodily metabolism in its relations with wider metabolic changes—that is, the exchange of matter among human society and nature that Marx called "social metabolism." We discuss these changes further in chapter 3, which explores the making of the protein powder industry. For now, suffice it to say that the bodily labor required to metabolize protein, and the work required to mass produce it, can be thought together when protein is understood as socioecological "all the way down."

Bringing these insights together, we can offer some conceptual dimensions of protein that depart from the reductive or decontextualized notions we have highlighted en route. We contend that protein is best understood as an elusive, dynamic, contingent, multiplicitous force: *elusive* because two centuries of proliferating investigation has repeatedly found protein to evade simple definition or categorization; *dynamic* because the agency of protein, its capacity to act and make a difference within bodies and wherever bodies and environments commingle, is increasingly recognized by biochemists, molecular biologists, ecologists, and social scientists alike, albeit with differing emphases and conclusions; *contingent* because claims about what protein is or does come with caveats about the involvement of other chemical bodies, or scientific apparatus, or the labor of humans (or animals), without which protein would not materialize as expected or "swerve" off course in ways that confound its followers and adherents; and *multiplicitous* because the forms it takes as muscle, food, waste, molecules, chemicals, and more must be understood not as entirely distinct or homogeneous but, in Annemarie Mol's terms, as versions of a phenomenon manifesting differently, those differences owing to a combination of its elusiveness, dynamism, and contingency.[60]

To these dimensions, we can add another, one that encapsulates our arguments about protein's social power. Protein is *adhesive* in that it can be (and has been) attached to an eclectic range of biopolitical imperatives, commercial ventures, and health and fitness goals, from global problems of population hunger and the climate crisis to everyday matters of meal preparation. Often its invocation is as a solution to or intervention in these outcomes, while at other times it serves as an underpinning rationale for social and political action, a unifying nutrient beyond reproach. Crucially, it is protein's capacities and tendencies that make its adherence to these problems and projects potent and enduring. In other words, were protein knowledge absolute, were its secrets revealed as a matter of record for all to survey, then its influence and status would surely diminish. As it stands, protein's power lies in its potential, an ambiguous glow of possibilities that stem from the elusiveness, dynamism, contingency, and multiplicity through which it is adhered to all manner of desires, bodies, and worlds.

Following Protein: Overview and Chapter Outline

To state plainly what is likely obvious by this point: Protein can be hard to pin down. We have pursued our subject for years now, like paparazzi in search of the next exclusive shot, of a unique glimpse into the life of a celebrity who is both familiar and elusive, everywhere and nowhere. Sometimes we'd get close, only for protein to change appearances, to transform into something else as it moved from one disciplinary domain, social location, or biological entity to another. Conventionally, the biochemists, biologists, agricultural and environmental scientists, nutritionists, exercise physiologists, and others who work on protein do so in silos, often operating with differing definitions of what it is, what it does, and what problems, if any, it presents. In the meantime, with a few notable exceptions, social scientists are absent from the conversation.[61] This book addresses this fragmentation by following its subject across disciplinary lines, paying particular attention to the generation and contestation of protein knowledge and the propensity of this disorderly biological agent to challenge established insights and refuse attempts to tame its capacities.[62] Where it has proven necessary to isolate or contain protein in a particular place, scene, or time, we have sought to retain a conceptualization of this complex substance as a "contentious synthesis" of humans and nonhumans, of "bodies becoming other bodies," and thus, in Eric Sarmiento's words, as a point of departure for questions of "embodiment, relationality, power, and collective becomings."[63] This approach highlights how human bodies are unevenly incorporated into commodity relations as producers and consumers of protein, highlighting the racial and colonial infrastructures that undergird protein booms past and present.

Through these theoretical and methodological considerations, we arrive at the overarching contention that protein's status as a nutritional superstar—its unimpeachable character—has little to do with most people's actual dietary needs. The science of protein intake is evolving and contested, and while other scholars seek greater certainty about protein's essence and function through the production of ever more precise forms of knowledge, we suggest that only a conjunctural analysis, one that takes seriously the multiple socio-ecological histories that have produced protein as a coherent category and valuable commodity, can adequately explain its enduring cultural significance. The tendency to take for granted protein's ontological complexity and to naturalize its premium status can be found even among the growing community of scholars producing critical work on protein, where the "why protein, why now" question is rarely broached. Seeking to join this conversation, we offer a story that holds together protein's vitality, its imbrication in relations of power, and its cultural appeal, historically and in the present. In attending to the consequences of protein primacy, we reveal how the strengthening of some bodies through protein-enriched diets occurs at the expense of others. Our argument emphasizes that the cultural preoccupation with protein is not innocent, in other words, and that its effects are felt because of—not despite—its elusive, dynamic, contingent, multiplicitous, and adhesive propensities.

The chapters that follow each offer historically situated accounts of specific moments, sites, and spaces in which attempts have been made to harness protein's meanings and material forms. As readers will see, these moments coincide with different phases and dimensions of capitalism—contingencies that we highlight as our analysis proceeds. We begin with the emergence of protein as a nutritional category and with our realization that debates about what protein is, and what the term *protein* should rightfully designate, are as long-standing as modern science itself. These debates form the focus of chapter 1, "What Is Protein and Why Does It Matter? Mystery and Magnetism from Molecules to Meat." German scientist and entrepreneur Justus von Liebig plays a prominent role in this narrative. We show how his 1842 proclamation of protein as the only "true" nutrient and meat as the optimal source of protein was crucial to securing protein's cultural ascendence, its insertion into frameworks of racial superiority, and its attachment to capitalist projects that include the making of a global food market. While Liebig helped establish protein's enduring valence, his findings were contested at every stage. As we trace these contestations, we show how this epistemological uncertainty has persisted into the twenty-first century and situate our work in relation to that of other social scientists who have written about the elusive and wily nature of proteins. Whereas that scholarship tends to focus on protein's biography as it unfolds in the research lab, we trace its manifestations and travels across broader disciplinary and social domains, establishing what can be learned by following protein through the multiple and diffuse political projects to which it has been made to adhere.[64]

In chapter 2, "The Great Protein Fiasco, Then and Now: Nutrition, Development, and the Trouble with Growth," we revisit one of nutrition history's most prominent conflicts. During the 1950s and 1960s, considerable scholarly and economic resources were invested in the subsequently disproven idea that there existed a deep and deadly "protein gap" between the world's rich and poor. Our analysis centers the ways that racial and colonial thinking about ideal corporeal growth and size, in concert with discourses of development and imperatives to address Western agricultural overproduction, helped propel the idea of a global protein crisis. We situate our analysis in the *longue durée* of protein history, arguing that the temporal and spatial persistence of protein fetishization linked to the growth of bodies, economies, and empires brings into question the limited scope connoted by understanding this period as a "fiasco." Our argument is not that protein deficiency or unequal access to it was (or is) a fiction. Rather, we contend that protein supplementation was mobilized as a universal, market-friendly, technocratic solution to a set of colonial social, agricultural, and labor conditions that were driving malnourishment. This is the story not of a mistake, in other words, but of a shape-shifting continuity by which protein knowledges help reproduce globally stratified and physically embodied dichotomies between healthy and unhealthy bodies, in concert with responsibilizing discourses of lifestyle and development.

In chapter 3, "From Gutter to Gold: A Political Ecology of the Protein Powder Industry," we start to provide answers to the question of what protein does by focusing on the development of the most popular protein supplement and additive: whey. Our analysis is set in mid-twentieth-century Canada and the United States, where the rapid industrialization of dairy agriculture and its ecological fallout played leading roles in the commodification of whey effluent. In the official storyline attached to whey's twentieth-century transformation, capital investment and technoscientific ingenuity responded as one to an environmental problem—the massive quantities of toxic whey effluent generated by intensive dairy production—and produced not only a solution but also a multibillion-dollar protein powder market. Our version of whey's biography diverges from this triumphant tale. In exploring the transmogrification of whey from a noxious by-product of dairy overproduction to a nutritional commodity deemed essential to many diet and fitness regimens, we find a multiplicitous substance with no simple line of ascendance but instead a messy ancestry of techno-innovations, capitalist imperatives, dairy cow exploitation, environmental campaigning, and a biopolitical shift toward the optimization of bodily health. Following whey powder through its historical emergence and into the formation of a protein powder market, we show how protein travels and mutates through and with bodies and machines in a dance of dispersion and assemblage that creates surpluses of value and waste, strengthening some bodies and compromising others. In so doing, we consider how protein resists and refuses attempts to harness its vitality for all sorts of ends. Whey powder—itself a technoscientific outcome born of attempts to manage the unruly and environmentally destructive proteins constitutive of whey waste—thereby represents a key site through which to assess the epistemological and political implications of protein's agentic capacities.

Having established whey powder as a primary vehicle for facilitating the mass production and diffusion of protein supplementation, we shift our focus to spaces of protein marketing and consumption. In chapter 4, "A Poverty of Flesh? Sarcopenia, Aging, and the Economization of Protein Deficiency," we trace the development of sarcopenia (loss of skeletal muscle mass and strength) as a category of disease that positions all humans as either currently or potentially deficient in protein. By situating leading research on sarcopenia in the conjunctural conditions of its emergence in the late 1980s, we show how sarcopenia became constituted as a disease of protein deficiency at a particular historical moment, one in which the need to find homes for nutritional sources of protein, the defunding of healthcare for aging populations, the frontier-building demands of the neoliberal university, and the ascendance of protein's status among nutricentric subjects coalesce. Through sarcopenia, protein deficiency is established as an inevitable reality of advancing years on earth, with consequences for how those years are experienced. But it also becomes a pliable, reversible phenomenon, something inexorable unless one undertakes certain practices of the body to mitigate its effects.

In chapter 5, "Protein in the Muscular Manosphere: Supplementation, Self-Optimization, and Microfascism in Men's Fitness Culture," we pursue amino acid supplementation into perhaps the most obvious site for analysis in a book about protein. Our focus is on what we call the "muscular manosphere"—a US-rooted, globally resonant, networked men's body culture where life optimization and self-ownership are fueled by protein powder and other lifestyle fixes for masculinities perceived as under threat and in need of revival. As we trace protein's commodified form through the lifestyle prescriptions and downline economy of some of the manosphere's most prominent entrepreneurs, we depart somewhat from the previous chapter. There we explored investment in the fit body as a quintessential manifestation of the self-responsible, productive, individuated, neoliberal subject. Here we propose a more complex relationship with neoliberalism's tenets, one that can emerge only as neoliberalism flounders. We argue that in the face of climate and labor precarity, existential insecurity, and a growing distrust of market forces, protein-fueled training of the body becomes a salve for a wounded Trumpian masculinity with links to a latent and emergent "microfascism" in which misogyny and whiteness are inherent to the absolute focus on bodily control and sovereign, optimized codes for living.

Our epilogue takes the long history of protein supplementation told in the previous chapters as a point of departure for confronting a number of key tensions in the present. The backdrop for our discussion is the place of protein in the contemporary culture wars, its adhesion to meat-first populism, and the existential threat that protein innovation is felt to pose. We offer reflections on how and why the protein supplementation market continues to expand even as the science underpinning it remains contested, even as its manifestations old and new are challenged by competing visions of climate and sustainability, and even as answers to the seemingly simple dietary questions "How much?" and "What kind?" become more desired and harder to find.

1

WHAT IS PROTEIN AND WHY
DOES IT MATTER?

Mystery and Magnetism from Molecules to Meat

The result of every such investigation, if it bear in any degree the stamp of perfection, may always be given in a few words; but these few words are eternal truths, to the discovery of which numberless experiments were essential. The researches themselves, the laborious experiments and complicated apparatus, are forgotten as soon as the truth is ascertained. —Justus von Liebig, "Animal Chemistry" (1842)

What comes to mind when you hear the word *protein*? A chunk of chicken, beef, or tofu that gets added to your meal bowl? A scoop of dehydrated whey powder blended into your postworkout shake? A nutritional category on the label of your can of tuna? A type of food that must be consumed at every meal? The secret to muscular embodiment? Meat and its metonymic links to masculinity? The key to the healthfulness—or harmfulness—of many diets, in the West and elsewhere? The stuff of which your bones, muscles, and skin are composed? A sequence of amino acid molecules that does important work in the body? A key mechanism in the transmission and prevention of COVID-19 and other viral diseases? Protein is all of these things and more. Biological and social, medical and cultural, abstract and fleshy, singular and plural, sub-ject and object, human and nonhuman, animal and vegetable, poison and pan-acea, dead and alive, microscopic and the largest steak on the menu, protein is a dynamic and multiplicitous substance that evades easy classification. Yet it is also the primary nutritional category in contemporary dietary culture, one that presents a unified front belying its complex history, contested character, and entanglement in enduring hierarchies and systems of power.

What, then, is protein? How has its "thingness" been defined? From where do its cultural connotations emerge? Organized in response to these questions, this chapter has two main goals: first, to give readers a sense of what protein is, ontologically; and second, to trace the germination of some of the key scientific and social ideas that have been attached to protein, allowing its cultural resonance and entanglement with injurious hierarchies to per-sist over two centuries. This is no straightforward task. The development of Western protein science has, from the outset, been mired in conflict and uncertainty, and the path to consensus about the matter at its center is far from linear or cumulative. While this is true of all scientific inquiry, the his-tory of protein science is especially disorderly. In his work on the evolution of protein as a scientific concept, philosopher and historian of science Eduard Glas points to the diversity of experimental approaches, research questions, theoretical models, methodologies, forms of evidence, and underlying as-sumptions that have characterized scholarship in the area.[1] To this we would add that the messiness of protein science is in large part a reflection of what we now know, which is that there are millions if not billions of proteins found in nature (they are still being identified), that each protein is distinct in its structure and function, and that proteins are involved in almost every cel-lular process, making their study ubiquitous across the numerous fields and interests that comprise the biomedical and life sciences.

A thorough accounting of the development of protein science would thus be impossible and likely undesirable if one wanted to address the topic with any depth or specificity. The implication for our analysis is that while its broad historical and geographical scope reflects protein's convoluted biography, what follows is not a universal or comprehensive story of the development of our subject's characteristics and meanings, or a unifying or exhaustive ac-count of its variety of manifestations, the profusion of knowledges that seek to chart its behaviors, or the diversity of meanings attached to it. Rather, it is an account of a number of key moments in the effort to establish protein's ontological parameters, biochemical effects, and social valences, beginning with the emergence of Western biochemical science in nineteenth-century

FIGURE 1.1. What is protein? Illustration created by Shawn Forde.

Europe when German biochemist and entrepreneur Justus von Liebig designated protein the only "true" nutrient and arriving at contemporary theorizations of our subject.[2] As the chapter proceeds, we move from a focus on scientific debates about protein's substance and character, to a focus on histories of protein prospecting, commercialization, and meat-eating, with a specific focus on India under British rule. We do this to emphasize the racial and colonial roots of protein's ontology. At first glance, our attempt to get specific about the material composition and function of proteins in the early part of the chapter might seem at odds with our exploration of the socially constructed and contingent nature of knowledge about these attributes that forms the focus of the second half. But we view these challenges as closely linked. Through our analysis it becomes clear that there is no straightforward answer to the question "What is protein?," and that rather than obfuscating or impeding protein's ascension to über-nutrient status, this elusiveness and instability are crucial to its enduring position.

In starting the chapter with the "what" of protein, we join the handful of humanities and social science scholars who have written about its complex ontologies and the challenge this complexity has presented to researchers past and present. The writings of anthropologist Natasha Myers and political theorist Samantha Frost represent especially useful points of departure for our discussion, though the scope of their work and the questions they ask are quite different from ours.[3] Their primary focus is on *proteins* within the context of molecular, laboratory-based, natural science research, rather than *protein* within the context of food and nutrition, where our emphasis ultimately lies. These authors' contributions to knowledge are distinguished by their creative methodological approaches as they attempt to come to grips with the evasive, dynamic, and multiplicitous character of proteins. This same realization has proven central to our own project and impelled us to link the molecular to the dietary in order to even begin to adequately grapple with protein's scientific and social significance.

Frost's *Biocultural Creatures* emerges from eighteen months of coursework in the life sciences, which she undertook with the goal of elaborating the "basis for a new theory of the human."[4] Through studies of carbon, membranes, protein, oxygen, and temporality in organism-environment relationships, Frost reveals how every aspect of living bodies, from atom to protein to cell to organ, is mutually constituted through the interaction of body and environment, thus dissolving any notion of a distinct boundary between the human and the nonhuman, the social and the biological, the self and its context. Her explanation of how biochemical processes can unfold in "reliably repeated" ways without there being a master puppeteer coordinating their activity behind the scenes is especially useful for understanding the "agency" of proteins at the molecular level.[5] Proteins, she argues, have "direction without intention" as they respond to the constraints of their dynamic material environments and in turn create the conditions for further responses. Also pertinent to our analysis is Frost's argument that the body is best understood as "energy-in-transition rather than substance composed," an observation that highlights the permeability of bodily boundaries and the importance of a focus on action, movement, and process in understanding the ontology of protein and its function in the composition of humans and other "biocultural creatures."[6]

While Frost focuses on what an understanding of the biocultural quality of proteins might contribute to theorizations of the human, Myers is interested in how knowledge about protein is generated with and by humans. *Rendering Life Molecular: Models, Modelers, and Excitable Matter* is the result of five years of ethnographic fieldwork among structural biologists and biological engineers working in US research laboratories, and the affective, kinesthetic, embodied techniques they use to represent and understand the structure and function of proteins.[7] Such an approach is necessary because proteins exist beyond our unaided perception: Even with the most powerful microscopes available, we cannot see an individual protein molecule. Moreover, contrary to some textbook characterizations of protein science as a settled matter, knowledge about proteins is far from complete, and proteins themselves are neither predictable nor obedient. To understand proteins requires more than pure observation; it requires an epistemology of relating that Myers likens to choreography, as protein modelers use their bodies, senses, and intuition to come to grips with the constitution and activities of their "wily" subjects.[8]

Both scholars focus on protein in its cellular manifestations and use it as a descriptor for a large group of diverse molecular substances that have in common their amino acid constitution, but both also touch briefly on protein's manifestation as a nutrient and its use as a singular noun to describe a component of the diet. Frost appears to view these as different categories of things, writing in the introduction to her chapter on protein,

> Before my leap into biology, I had, generally speaking, a dietary conception of protein. That is, when I thought of protein, I thought in terms of what we eat: Muscles from some unfortunate critter, an element of nuts or dairy goods that make them an important food source, or a byproduct of a fermentation process through which beans or other vegetables are transformed into highly nutritious edibles like tofu. But proteins do not exist as bare chunks of protein substance located here and there in animal bodies or foods. Rather . . . proteins are variously sized molecules that do all kinds of work within and between cells.[9]

Myers also uses food as an entry point into her discussion of the meaning of proteins—in this case as they manifest in egg whites—but does so in a way that connects their molecular composition, which lies "beyond the limits of our perception," to the "slippery, translucent mass" that are the egg whites we might slide into a frying pan or whip up into a meringue.[10] Egg whites, she notes, are mostly water (88 percent), but about 10 percent of their weight is composed of protein, with one particular protein, ovalbumin, accounting for just over half of all their proteinous content.[11] Myers doesn't take her discussion of protein as food any further than this, but she does allow readers to see the material and discursive links between the different meanings of protein as they travel through different spheres, in this case, the scientific and the culinary.

We share with Frost and Myers a sensibility attuned to the dynamism of protein, to the labor it performs, and to the labor that must be performed by humans to render protein "visible, tangible, and workable at the molecular scale."[12] Our study breaches the confines of the lab, and social studies of twenty-first-century natural science, however, to explore the ontology and epistemology of protein across multiple, interconnected spheres of life, with a particular emphasis on its manifestation as a nutrient. That protein evades simple categorization and that its capricious nature is key to its social power is discernible from at least the mid-nineteenth century, when foundational debates in modern Western science were being played out, and terms to describe newly identified biochemical substances, proteins included, were being coined.

Starting our analysis in the 1830s reveals that as much as protein is having a big cultural moment in the 2020s, the current boom is only one iteration of its much longer biography as a "charismatic nutrient"—sociologist Aya Hirata Kimura's term for dietary substances that come to "command center stage" in food and nutritional politics at particular moments in history.[13] They do so not because of their inherent biochemical value, Kimura argues, but because of a conjuncture of social relations that mobilize them as simple yet compelling solutions to complex problems.[14] We can see this process at work in the circumstances through which protein was first imbued with its "magnetism"—the focus of the second part of the chapter. Liebig's scholarship, alongside his commercial endeavors to make a global business out of protein supplementation geared to strengthening European populations, is particularly key in this respect, and it is here that we start to find clues to the questions driving not just this chapter but the book as a whole: Why is it that protein is understood as healthy in and of itself? What conjuncture of relations helped endow this particular nutritional category with its singular power? What economic and social problems is protein thought to solve, and what problems does its status as a nutritional superstar generate?

Liebig wasn't the only researcher of his time promoting reductionist models of the living world, but he was an especially entrepreneurial and influential one. Frequently described as a "propagandist,"[15] Liebig assiduously chased the limelight, but he did so not just for his own satisfaction, or to finance the development of his growing "gentleman's" estate, but with the goal of extending the boundaries of chemistry to demonstrate its social and economic significance.[16] While Liebig's status as a "chemical gatekeeper" and father of protein science is widely recognized,[17] reading his contributions with a critical, contextual eye reveals that protein science and industry was, from its inception, enmeshed in concerns about labor power, economic productivity, and food availability, and implicated in the transformation and appropriation of land, resources, and work on which the success of industrial capitalism depended. Tracing these connections allows us to show, here and in the chapters that follow, how protein knowledge and production have emerged as mechanisms for attaching different kinds of value to different kinds of bodies through discourses of strength, deficiency, and supplementation tied to class, race, gender, and age orders—legacies that endure in one form or another today.

As we map the desires, systems, and modes of thought that Liebig's work helped generate, we suggest that he was a father not just of protein science but also of an extractivist global food system and of nutritionism, if not fixated and disordered eating. These roles are closely linked: As protein prospectors such as Liebig looked outward, seeking nitrogenous food imports from afar that could feed a growing European population, they also inspired a turning inward, toward an engagement with the body as a biochemical project requiring an appropriate balance of nutrients obtained through an efficacious dietary regimen. In the process, the nitrogen-rich foods that eventually came to be known as protein were installed as the fulcrum on which the notion of nutritional balance would rest, the messy politics on which protein's character is founded occasionally destabilizing but never displacing its hallowed status. Concomitantly, meat-eating was incorporated into logics of racial superiority (corporeal and cognitive), especially in India, and bodies, proteins, nutrition, and health were gradually divorced from the food system and from their larger social and ecological contexts—an ongoing, violent abstraction that we seek to make material in the pages that follow. The chapter thus lays the groundwork for understanding three key processes: how multiplicitous, proteinous matter has congealed, ontologically, as a preeminent nutritional category; how protein supplementation has emerged as an indispensable, legitimate practice; and how protein prescription and consumption are bound up with prevailing social hierarchies.

On "Baffling" and "Unruly" Substances: From Vagueness and (Un)Certainty to Dynamism and Multiplicity in Protein Science

The formative years of biochemical research in Western Europe were marked by an often-vicious dispute between Liebig and his Dutch rival Gerardus Johannes Mulder about the atomic composition and function of "animal substance," or what we now think of as proteinous matter.[18] In 1838, upon the recommendation of his Swedish colleague Jöns Jacob Berzelius, Mulder had introduced the word *protein* into the scientific literature. Mulder used the term to designate a complex grouping of carbon, hydrogen, nitrogen, and oxygen atoms combined with one or more sulfur and phosphorous atoms that were common to substances such as albumin (found in eggs and other animal products) and casein (found in mammalian milk). Mulder considered this grouping the "fundamental" element in nutrition, hence his choice of a descriptor derived from the Greek adjective *proteios*, meaning "primary" or "holding the first place."[19] It seems appropriate that the Greek god Proteus, known for his capacity to change himself into different shapes—that is, to be *protean*—should share an etymology with protein. Proteus's appellation was probably chosen because he was the first son of Poseidon. But protein historian Kenneth Carpenter suggests that Proteus's mutable qualities were on Mulder's mind when he named his molecule and referred in his explanation to the way that plants convert protein to make it edible for consumption by herbivores, who in turn convert protein to make it edible for carnivores.[20] For Mulder, then, protein was a versatile substance but one that had a unique and consistent elementary structure and constitution. Convinced of the significance of his findings, and foreshadowing contemporary popular usage, Mulder heralded protein as "the foodstuff of the whole animal kingdom"—a descriptor that endures to this day even though the knowledge underpinning it has shifted considerably.[21]

While Liebig shared Mulder's enthusiasm for the role of protein-like substances in human metabolism, he was ultimately unable to reconcile his findings with those that had led Mulder to coin the term *protein*. Liebig preferred to understand protein as a general descriptor for a "whole class of related substances, which shared the same atomic 'composition' . . . but not the same atomic 'arrangement.'"[22] In other words, Liebig agreed with Mulder about the consistent ingredients that constituted what we now call protein, but postulated that it was an oversimplification of protein structure to suggest that these ingredients remained stable in relation to one another as physiological processes unfold. According to biochemist and historian Hubert Vickery, after several years of intensive study, Liebig lamented in an 1845 letter that, although it was a pity, "no such thing as 'protein' exists."[23] Vickery goes on to note that Liebig's correspondent, Friedrich Wöhler, who was attempting to act as a peacemaker between the two men, claimed in response that "science had been enriched by the useful and appropriate word protein the meaning of which . . . was sufficiently vague."[24]

The looser meaning of the term persisted as the idea of a singular protein radical disappeared and the word *protein* gradually came to refer to the varieties of nitrogenous matter that were thought to contain body-building properties.[25] So, while Wöhler presaged the popular usage we see today in his attempt to orchestrate some shared amity, Liebig's instincts were more aligned with contemporary biochemical knowledge in which proteins are understood as infinitely variable. As Vickery noted in a 1942 paper, it would take another seventy years before the role of amino acids in converting proteins in food into proteins in animals became evident, thus upsetting any notion of a stable "protein" molecule and replacing it with the idea of variable chains of amino acids. Still, as Vickery writes, "today . . . we recognize the grain of truth" in Liebig's claim that no such thing as protein exists.[26] And while Mulder was incorrect, as Liebig thought, in his belief that there were a small number of proteins, we can see in Mulder's theory, and his selection of terminology, the seeds of an amino-acid-centric conceptualization of protein and the roots of the now-hegemonic view of protein as the "building block of life." Not until the first decade of the twentieth century, with the emergence in 1902 of the groundbreaking hypothesis positing that proteins consist of chains of amino acids linked by peptide bonds, would scientists come to see that both Mulder and Liebig were on the right track, but constrained by the available experimental resources and accompanying range of questions that were possible to ask at that time.[27]

New technologies and a growing body of knowledge went only so far toward clearing the muddy waters of protein science and typology. According to Glas, "Lacking a generally agreed upon frame of reference, physiological chemistry in the second half of the nineteenth century embarked on a variety of divergent lines of experimental inquiry, unconstrained by any leading theory or theories. Ongoing efforts to characterize proteins . . . did not lead to stable results."[28] The divergent lines of inquiry to which Glas refers included research on hemoglobin that moved conceptualizations of protein in the opposite direction from Mulder's proposition of a single common molecule.[29] With blood from humans and a range of other animals easily available for experimental purposes, and following the discovery of hemoglobin's crystalline construction in 1840, researchers were able to establish that hemoglobin was a protein and that its structures differed widely across a range of species.[30] In the words of biochemists and historians Charles Tanford and Jacqueline Reynolds, this discovery "conjured up vistas of a virtual infinity of proteins—thousands of unique species modifications superimposed on an ever-increasing variety of chemical and functional distinctions."[31] However, as the authors also note, the crystalline morphology of proteins is only vaguely related to "molecular details," and this research didn't yield much insight into the processes through which proteins are made or how they evolve.[32] Tanford and Reynolds go on to write, with reference to hemoglobin, that "nothing like this wealth of knowledge exists from that period for any other protein," including myosin, which was discovered in the same era but would not be understood at a molecular level for close to another century, the mechanisms of muscle contraction and locomotion in which it plays a key role remaining unknown through the corresponding period.[33]

According to Glas, "The emergence of a viable theory of protein constitution had to wait till the end of the nineteenth century, when renewed attempts were made to characterize protein building-blocks on the basis of sulphur contents and salt-binding capacities."[34] By this time, Mulder's earlier work along these lines had "sunk completely into oblivion," and Liebig's theories had lost "much of their credibility."[35] Nonetheless, historians of this period tend to agree that Liebig's early investment in the unique importance of proteinous matter never fully lost its momentum. As Tanford and Reynolds put it, "The very existence of the Liebig-Mulder dispute enhanced the status of proteins."[36]

Despite continued investment in the nutritional primacy of proteinous matter, the "sufficient vagueness" to which Wöhler had referred fifty-nine years earlier continued to trouble the scientific community, even as more feasible conceptualizations of its form and function emerged. The 1907 proceedings of the Physiological Society and the Chemical Society, scholarly organizations based in London, England, reveal a membership seeking to navigate a path through the confusion that arose from the "ambiguity of meaning" around protein.[37] In a four-page paper, "Protein Nomenclature," that resulted from the work of a joint committee, less emphasis is placed on the kinds of metaphysical and scientific disagreements about the character of protein in which Mulder and Liebig were embroiled and more emphasis on standardizing terminology in areas of study in which, to their minds at least, there was some scientific consensus. Translation was thus a concern for the committee, with the emergence of English-language hegemony in science writing seemingly motivating their desire to move beyond the mix of Latin, English, German, and French elements and words that had been used to refer to protein-like substances until that point. To that end, the committee noted, "It is well known that much confusion arises at the present time from the lack of any understanding, either here or abroad, as to the exact sense in which the various names applied to proteins (Proteinstoffe, Eiweisskörper, Albuminoids) and their derivatives shall be used. No little difficulty is created by the use of a term in different senses, as well as ambiguity of meaning in some cases."[38] Calling for "some degree of uniformity," though recognizing that "only provisional recommendations are desirable in the existing state of knowledge," the committee recommended that "the word Proteid . . . should be abolished," and "albuminoid," which was in common usage at the time, "if used at all . . . should be regarded as a synonym of protein."[39]

Despite the definitive language used in this passage, the committee's understanding of the meaning of protein remained at once vague and complex. "The word Protein is recommended as the general name of the whole group of substances under consideration," they wrote, going on to list names for seven "sub-classes" of protein.[40] These included substances such as protamines, which were described as "simple members of the group," and histones, which were described as "more complex substances" that "probably pass gradually" into protamines and vice versa.[41] It is notable that the term *amino acid* only appears once in the document, perhaps reflecting the committee's concern with categorizing "sub-classes" of proteins rather than their molecular composition, but its absence is striking given the arrival of the peptide hypothesis five years prior and the emergence of amino acids as central to standard definitions of protein in that era.[42] A somewhat tentative and convoluted tone characterizes the outstanding pages of the document, highlighting how troublesome the categorization of protein remained for this group of experts, even though their recommended nomenclature for the general type of matter in question has endured to the present.

While the general name for this category of substance may have stuck, the stretchiness of protein as a linguistic category used to designate substances ranging from molecules invisible to the naked eye to oversized chicken breasts persists. This ambiguity reflects a complex etymological history across multiple languages and periods of time, but it also indicates the contentiousness of knowledge about the substance, structure, and function of *protein*—now mostly used as a singular noun to refer to a macronutrient or component of the diet—or *proteins*—now mostly used to refer to a variety of molecular substances which, among many other characteristics and functions, constitute, biochemically, a portion of the former. Indeed, science about protein has remained in perpetual motion through the twentieth and twenty-first centuries. A 1931 article, "The History of the Discovery of the Amino Acids," authored by prominent biochemists Hubert Vickery and Carl Schmidt, notes that while there had been "rapid" progress in protein science in recent years, "the theory of protein constitution must be brought to a far more highly developed state than it is at present, before we shall have much reason to be satisfied with our knowledge of these baffling substances."[43]

The twentieth century saw the development of a plethora of techniques and technologies that would significantly advance the field of protein science, including chromatography, X-ray crystallography, nuclear magnetic resonance spectroscopy, and computational prediction, which together have provided deeper insight into the structure of proteins, particularly the precise composition and sequence of amino acids along a polypeptide chain, and the specific mechanisms involved in the crucial matter of protein folding. But many mysteries remain. In 2022, ninety years after Vickery and Schmidt noted the "problems for investigation" proteins provide, science journalist Laura Howes, writing about the effort to map the entire set of proteins in the human body, described the state of knowledge thus: "By some measures, scientists have nearly finished quantifying the human proteome. But by others, they have only just scratched the surface of the variety and complexity of our bodies' proteins. They have yet to find proteins for some known genes, and they haven't characterized the modified versions of some proteins. For other proteins, scientists don't know what they look like or what they do in cells."[44]

At one level, this is just how science works. Ideas and descriptors change and multiply as inquiry unfolds. But this process is intensified in the siloed world of contemporary science, where research occurs across a large range of highly specialized disciplinary formations, making objects of knowledge especially mutable. Organic chemistry, biochemistry, crystallography, molecular biology, nutritional science, agricultural engineering, environmental science, kinesiology, anthropology, geography, food studies, political theory, sociology, history, and philosophy are just some of the areas of study we have encountered in our pursuit of protein. While some of these fields, or corners of these fields, share in common a focus on one particular version of what protein is, others have their own version or versions at stake. Where one scientist's protein might be a microscopic crystal, another scientist's protein might be an amino acid sequence, oncogene, enzyme, macronutrient, powder supplement, or handful of almonds. As Annemarie Mol notes in her ground-breaking work on multiplicity, these versions may be strongly or tenuously linked, consistent or in tension with one another. Even within versions, the typology, ontology, structure, and function of proteins is far from settled.[45]

Out of this multiplicity, there are four unquestionable points of consensus about proteins today: Proteins are composed of amino acids; they perform pivotal functions in almost all biological processes, including nutrition; how they "fold" influences how they act; and they are inordinately complicated if not mysterious elements of the biocultural world. More specifically, it is now understood that each individual protein is composed of some combination of twenty amino acids (twenty-two if one includes the two rare amino acids that require a special mechanism to be incorporated into a protein) and that the atoms most prevalent in amino acids are carbon, hydrogen, nitrogen, oxygen, and sulfur. With regard to protein as nutrition, specifically, we understand that the nitrogen humans need to survive enters the body through amino acids, that they derive nine "indispensable" or "essential" amino acids from the food that they eat, and that they have the capacity to make eleven acids themselves. We also know that when humans ingest food, their bodies break the proteins down into individual amino acids and then remake them into the proteins they need for all manner of bodily tasks.[46] And we know that proteins are among the most abundant molecules in living systems (the average cell contains 42 million proteins) and are commonly labeled "the agents of biological function."[47]

As Frost writes, in the "frenetically active cellular world, thousands of proteins do innumerable tasks."[48] Using the analogy of cityscapes dense with traffic to envisage proteinous activity, Frost explains how proteins give structure and shape to cells, and anchor, stretch, retract, or bind them in support of growth, development, and movement. Proteins also act as enzymes that catalyze and regulate the thousands of chemical reactions that compose cellular life, as messengers that transmit signals and coordinate biological processes among different actors and substances, as vehicles that transport chemicals within and beyond the cell over a variety of distances, and as antibodies that bar entry to bacteria or viruses or combine with them to store or repackage them for future, beneficial use. The exact role any particular protein plays is understood to depend on the precise sequence of the amino acids it is composed of, the way these amino acids interact with one another, and the shape they take—that is, how they "fold."

It is here that things start to get less certain, or rather where uncertainty and contingency become less definitional sticking points and more defining characteristics of protein. For an amino acid chain, or polypeptide, to carry out its task, it must bend itself into particular configurations, but the form that any specific configuration takes depends on a plethora of other chemical interactions and contortions that occur among various side chains or secondary formations. And these in turn are dependent for their composition and structure on the presence of other polypeptides and the broader chemical environment, including its water content. Philosopher of science Joyce Havstad explains that protein scientists manage this "incredible complexity . . . (somewhat) by dividing protein structure into a hierarchy of four levels": the primary structure, which is the sequence of amino acids; the secondary structure, which is the initial folding of the sequence into shapes like helices and sheets; the tertiary structure, which arranges or folds the secondary structures to form a three-dimensional, often globular, shape; and the quaternary structure, which, unlike the other three levels, is not common to all proteins and only arises when two or more polypeptide chains with their own primary, secondary, and tertiary structures are packed together to form another, "complete" protein.[49]

Havstad's use of the word *somewhat* suggests that the complexity of protein cannot be fully captured by this taxonomy, a point Myers illustrates this way: "Amino acid sequences do not communicate the full information required for a protein to acquire its active form. For example, many different amino acid sequences can produce similar tertiary structures, and highly similar sequences can produce different folds. Moreover, some proteins can fold only in the presence of other molecules called chaperones, which support these proteins in achieving their final form." Thus, Myers concludes, "protein folding is a complex and poorly understood process."[50] This is why protein crystallographers and other modelers must deploy creative, intuitive, and embodied strategies to figure out the structures of these tiny, "unruly," fast-moving substances that evade detection by even the most powerful microscopes or computers.[51] Writing in 2016, Havstad notes that despite the "incredible time and energy" researchers invest in producing their models, "out of the estimated 30 to 70 million proteins currently known to science, less than 100,000 proteins have structures that have been 'solved' by scientists."[52] Since then, the number of solutions has increased to an estimated 194,000, but the number of proteins has expanded with it—to an estimated 200 million. As Havstad explains, "Even the structure of 'small' proteins with sequences of 'only' a few hundred amino acids can still, despite many recent advances in technology, be incredibly hard to solve because of the myriad contortions, relations, shapes, turns, and complexes which these macromolecules can form."[53] Knowledge about their roles is also partial, with "the function of 40% of the proteins encoded by the human genome . . . unknown," according to Garrett, author of a best-selling biochemistry textbook.[54] The fact that the approximate range of "currently known" proteins is so large, and that so much about their roles is undetermined, underscores the evasiveness, dexterity, and complexity that characterize our subject.

Some important advancements have occurred in recent years, including the development of potentially more cost- and time-efficient ways to predict how some proteins take shape and function. As science journalist Rowan Jacobsen writes, "New insights and breakthroughs in artificial intelligence are coaxing, or forcing, proteins to give up their secrets."[55] The most prominent AI system for predicting protein shapes, Google-owned AlphaFold, has provoked great excitement with its release of predictions for 200 million protein structures, which the company makes available via a public database.[56] While researchers are using the predictions to gain new insights into all manner of pressing challenges, from vaccines to honeybee health, it has many limitations, including that it produces predictions, not "true shapes." As protein biophysicist Julia Forman-Kay notes, "Painstaking experiments remain crucial to understanding how proteins fold."[57] In the meantime, many aspects of what scientists describe as the "mystery" of protein remain out of reach.

The Only "True" Nutrient

Up to this point, our discussion has been focused on the nomenclature and ontology of protein, rather than its social functions and cultural reputations. To begin to broach the latter, and to understand the implications of its manifestation in nutrient form, we must return to Liebig and his colleagues. With hindsight, it is now evident that the struggles of Mulder and Liebig to name and understand their subject were related to the fact that *protein* as we use the term today "denotes a subject matter which itself shows various nested or superposed levels of structure and organization," which are "absolutely essential," Glas emphasizes, for comprehending its biological activity.[58] With only very partial knowledge about molecular and supramolecular composition and constitution, and lacking the tools to more directly access these levels of structure and the relationships among them, Mulder's and Liebig's hypotheses could only go so far. Protein's particular complexity in comparison with other molecular forms was also undoubtedly at play. As Glas writes,

"Whereas the structural formula of most organic molecules tells almost the whole story about their possible spatial conformations and properties," this is not true of protein, as the preceding discussion makes clear.[59] It is in reference to this ontological difference that Glas makes a rare gesture to the realm of nutrition, noting that beyond protein's layered structure, the "law of definite proportions," whereby every chemical compound contains fixed and constant proportions of its constituent elements, does not apply to proteins.[60] This is in contrast, he writes, to carbohydrates and fats, which are "classes of identical molecules of definite elementary composition," and, unlike proteins, do not vary between species or individual organisms.[61] Viewed in this light, we can begin to see how the uniqueness of this thing we now call protein, its complexity and opacity, even in its stripped-down molecular form, might have fed into its broader cultural significance—the problem to which we now turn.

The identification of nitrogen at the end of the eighteenth century and the development of techniques to measure its levels had enabled researchers to accelerate their endeavors to understand plant growth, soil chemistry, and other determinants of robust agricultural production and a well-fed European population.[62] In an 1843 work, French chemist Jean-Baptiste Boussingault, one of Liebig's key interlocuters, noted that farm animal feeds that were high in nitrogen tended to have a higher overall nutritive value. The conclusion Boussingault drew from this observation helped set in motion a vision of nitrogenous foods that persists to this day: "The nutritious principle of plants and their products resides in their nitrogen-containing principles and consequently their nutritious powers are in proportion to the nitrogen that they contain. . . . Although nitrogen-containing principles are not alone sufficient for the nutrition of animals they are the limiting factors in all kinds of vegetable foods."[63] In other passages, Boussingault argued that the nutritional quality of a "vegetable substance" is proportional to its gluten and albumin content, or what he variously called "animal substances," "flesh," or "nitrogen-containing principles." In this way, Boussingault helped move proteinous or meaty matter to center stage.[64]

Boussingault's research was primarily based on feeding experiments with farm animals, but Liebig undertook his work through the lens of what we now call organic chemistry, or the level of the compounds that constitute living things. Like other scientists of the era, Liebig was trying to answer fundamental questions about the stuff from which bodies are made, the processes through which food becomes flesh, and the mechanisms influencing corporeal composition, size, strength, and stamina.[65] Nutrition was thus a central focus of his research, which was not the case for many of the protein scientists who succeeded him, given the growth and diversification of the field thereafter.

Liebig's analysis of muscle fibers from a variety of animals had failed to reveal the presence of carbohydrate or fat, leading him to assume that protein, or the nitrogenous compounds that fell into that category, provided both the substance of muscles and the energy required to fuel their work through an "explosive breakdown of the protein molecules themselves."[66] In one crucial experiment, Liebig found that the muscles of foxes "killed in the chase" contained ten times more creatine than a captive fox "fed on flesh for two hundred days."[67] Based on these observations about creatine, an amino acid that occurs naturally in the body and is named from the Greek word for "meat" or "flesh," Liebig concluded that proteinous compounds were alone responsible for building and replenishing tissue, providing the energy for muscles to contract, and allowing humans to move and survive. Concomitantly, Liebig relegated the function of carbohydrates and fats to their role in respiration and heat, leaving protein with the only genuine nutritive role. Liebig understood, correctly, that all protein is plant-based at its origin, and only becomes "meatified" as the animals we consume, consume it. But because the muscles of the physically active and strong animals Liebig studied were composed of protein, and because he thought that more work was necessary to convert plant proteins into flesh, he understood animal proteins—more specifically, meat—as the most important and efficient source of nutrition. And because he thought muscles could be restored only through proteinous food, he helped establish the idea that more protein was needed to supplement the "normal" diet for bodies—or at least the European male bodies he was primarily concerned with—that were moving and at work.

Many of Liebig's ideas have been subsequently refined or rejected, including his assertions about the role of protein in energy production and muscular activity, but some of his claims are not entirely misaligned with current wisdom. Proteinous foods *can* be used to form highly prized muscular physiques when accompanied by appropriate amounts of exertion. Protein in muscle *can* be used as a source of fuel for underfed bodies or in the latter stages of endurance exercise, even though under ordinary conditions it meets only about 5 percent of the body's energy needs and cannot be stored like glycogen or fat. And creatine, which occurs naturally in the body but is also derived from red meat, seafood, and synthetic supplements, is *thought* to help supply energy and bulk to working muscles.[68] These partial insights aside, critics tend to agree with Carpenter's assessment that Liebig's ideas about protein "were really no more than speculative hypotheses" and that "the scientific bases for them were to disintegrate quite quickly."[69] Indeed, by 1870, all but Liebig's most ardent followers had accepted that protein was not the main or obligatory source of energy.[70] While the primary function of carbohydrates in fueling the body would be settled, if not fully understood, before the close of the nineteenth century, to this day, and despite some milestone clarifications, knowledge of how muscles contract and of the role of proteins within this process remains "far from complete," with muscle properties and functions in human movement "largely unknown."[71] Nonetheless, the pronouncement of Liebig's theories "as dogma," and their general acknowledgment as such, meant that they became "a generally accepted pattern of ideas . . . to which succeeding observations had to be fitted."[72] In other words, Liebig's protein-dominant, meat-first, supplement-oriented approach to diet became a paradigm in the Kuhnian sense, one that has been subject to small shifts over time but never entirely displaced.[73]

Nutritionism and the Persistence of Protein Primacy

We want to take a moment here to explore how and why Liebig's ideas about protein have endured. We do so not to overstate his individual genius or historical agency but to convey how his thought, and his dedication to the popularization of scientific knowledge, emerged in conjunction with a specific set of political and economic arrangements that functioned to entrench in protein's biography a particular set of cultural ideals and commitments.[74] Liebig scholars have suggested that one key vector for the dissemination of his ideas was the many graduate students from the United States, Britain, France, and Germany who came to train at his Giessen lab.[75] Attracted by a pedagogical approach that allowed them to conduct experiments themselves, rather than simply watching and listening to their mentor's demonstrations and lectures as was the convention at the time, these students would then return to their homelands, where they took up his ideas, including his gospel of protein, even if they often did so with conceptualizations that challenged Liebig's theories of its structure and function.[76] Also important was the evolution of Liebig's early interest in the chemical composition of organic substances and animal metabolism into a broader, socio-physiological concern with agricultural chemistry, food production, ideal dietary intake, domestic cookery, infant health, and other topics that facilitated an engagement with his ideas by a wider audience, something he worked with intention to generate.[77]

Liebig's efforts at self-promotion paid off, particularly in Britain, where participation in an emerging international culture of scientific exchange linked to the rise of industrial capitalism was especially enthusiastic.[78] Indeed, geographer Greta Marchesi argues that the broadening of Liebig's professional activities allowed him to emerge as the "gatekeeper between the industrializing world and the field of chemistry."[79] This role is evident in a much-read three-part series published in the *Lancet* in 1869 where Liebig expended thousands of words dissecting the nutritional needs of a wide array of figures, his propensity for keeping an "eye on the bigger picture" apparent from the opening paragraph, where he provided the following justification for the analysis that followed:

> It has been said that if man could live on air and water, there would be an end at once of the notions master and servant, sovereign and subject, friend and foe, hatred and affection, virtue and vice, right and wrong, &c., and that our political commonwealth, social and family life, our intercommunication, trade, commerce, and industry, art and science, in short, all that makes man what he is, would not be if he had not a stomach, and were not subjected to a natural law which obliges him daily to take a certain quantum of nourishment. It is therefore worthwhile to answer the question why in reality man eats and drinks, and what the substances are which, received in the body during a succession of years, have an influence on the duration of his life.[80]

Shifting seamlessly between discussions of the "component parts of human food" and "the fodder of animals," Liebig's inquiries moved from "a soldier, in a time of peace" and the "English navvies who were sent out during the Crimean war" to a variety of farm animals and a "bear kept at the Anatomical Museum of Giessen."[81] By the time of writing, Liebig's inquiries had forced him to acknowledge that "heat-giving substances" such as fat and starch, and "nutritive salts" such as magnesium and iron, were crucial for "nourishing and sustaining life," yet he maintained his belief in the "albuminates" or "flesh producers" as the material source of all components of the body, "animated and plastic," here and across the span of his career.[82]

Liebig was particularly concerned about the diets of working-class men in his country of origin, noting at one point that "the badly-fed German workman wants in England and America a month's diet abounding in albuminates before he is able to compete with the English or American workman."[83] In a typical passage, he contrasts in minute detail the food consumed in "seven months by 95 men in a Munich brewery" with that consumed by "a woodman in the Bavarian highlands," noting that the brewery workers "in meat alone, consumed 120 grammes of albuminates, with bread—altogether, from 160 to 170 grammes daily: thus nearly three-quarters meat, and one quarter bread," in contrast to the woodman, who consumed 130 grams of albuminates daily.[84] Liebig acknowledged that the woodsman's work was "hard," but claimed that it did not require much energy, since "after every blow with his axe he can rest as long as he pleases, for the tree stands still, and does not require him to make haste." Brewery work, in contrast, "is the hardest of all, and only strong men are able to endure it, for the operations follow one another uninterruptedly, and tax the strength of the workman unceasingly."[85] While Liebig was clearly attuned to the social context in which these men were performing their roles, and to how the rhythm of manual labor in industrial capitalist settings might differ from that in nonindustrialized, rural settings, he erroneously understood the distinctions in the consumption habits of the two groups to reflect the different kinds of physical labor they undertook, rather than their access to food—a common tenet of nutritional science well into the twentieth century. His argument was thus that the brewery worker, needing relatively fast access to the albuminous nutrients because of the pace of his work, required a meat diet, while for the woodsman, who could fell trees at his own pace, "a purely vegetable diet" would "suffice."[86]

Given the focus on paid labor, it's unsurprising that women and their work are largely absent in this series. The exception is a handful of passages where Liebig's long-term interest in infant feeding (he had patented his "Food for Babies" just four years prior) comes to the fore. Lamenting the widespread use of pap, a porridge often made with wheat and milk or water, he approvingly quotes physician John Zimmerman, who had written many decades earlier that it was "easier to move a mountain than to convince a brainless woman of the disadvantages of pap."[87]

Later in the series, starting with the supposition that the price of food does not necessarily align with its nutritive value, Liebig offered a long sequence of calculations designed to reveal the average price of one pound of albuminate in seven categories of animal and plant foods. His idea was that such work would lead to an understanding of "what mixture of substances containing most nutrition will cost least."[88] He concluded, among other things, that eggs are the most expensive among the common foods "furnished by animals" and cheese the least, and that albuminates in vegetables "must of necessity be much cheaper than meat" because otherwise it would be impossible to feed cattle.[89] While Liebig acknowledged the necessarily rough nature of these estimates, noting that "to these calculations no greater value must be attached than they deserve," he went on to say, "My aim will be attained if I have succeeded in convincing the reader that even when we eat we may do so according to fixed principles, and he who has learned to do so has learned something of the art of prolonging life."[90]

With a focus on quantitative analysis (no matter how approximate) and "fixed" dietary principles as the key to longevity, we see the nutritionist elements of Liebig's paradigm come to the fore. While Marchesi's analysis of Liebig's impact is focused on a different aspect of his research—that which famously influenced Karl Marx's writing on the extraction and depletion of soil nutrients as a key crisis of "social metabolism" under industrial capitalism—her analytic approach is useful for our purposes. Moreover, soil and protein chemistry are closely connected: Cultivation of the soil through agriculture is the primary source of food production in the modern world, and all animals ultimately obtain their protein from plants, most of which grow in soil, a point that Liebig himself made in his descriptions of the food chain.[91] Of particular interest for us is Marchesi's explanation of the "sustained utility" of Liebig's idea that "plant growth was reducible to measurable chemical interactions," and the insights that can be gleaned from her analysis regarding protein's ongoing primacy in the nutritional world. Marchesi argues that Liebig's interventions in the science of soil were both "practical and epistemological," noting that while new insights have "long eclipsed" his "technical recommendations, his model of soil as a chemical repository has had a stubborn appeal."[92] The parallels with the uneven legacy of Liebig's work with protein are striking and can be explained in part by the similar epistemological underpinnings of his research on both soil and protein. Marchesi writes that Liebig believed the following: "All material bodies could be productively reduced to a series of relationships between discrete chemical agents and interventions performed at the level of those chemicals. Furthermore, these agents could translate uniformly between dramatically different contexts, building blocks in a world of circulating equivalents in which all matter was nominally equal to the sum of its parts."[93] The implications of Liebig's epistemological perspective were profound: Readers will likely recognize its reflection in his formulaic thinking about the ideal diet, and by extension in the nutritionism that helps organize contemporary approaches to food both within and beyond the world of science. Furthermore, the idea that chemical elements were like mathematical variables or parts of an engine that could be manipulated and tinkered with opened up the possibility of "systematic, informed interventions" into metabolic processes, including, we would argue, the human diet.[94] While the technicalities of Liebig's theory of protein and its application to dietary interventions have long since been rejected, the elegance and portability of a nutritionist model with protein at its center, and the opportunities this model promised for the improvement of populations (or at least of the European working-class men who comprised his primary focus), have maintained their appeal to this day.

Let Them Eat Meat: Protein Prospecting and Colonial-Capitalist World-Making

There is more to this story than the appeal of a transferable scientific model, which, in the case of plant agriculture, had the effect of promoting standardized, industrial, chemical management across diverse soils and landscapes. To understand why Liebig's model of plant growth has persisted in the face of a huge body of contrary research stressing the importance of localized, organic approaches to soil biology and ecology, Marchesi turns to the historical context in which it unfolded. Again, the parallels and intersections with protein's history are notable. In Marchesi's view, Liebig's ideas would not have gained such traction without their entanglement in the mid-nineteenth-century transformation of capitalist agricultural production in which the scientist was invested. Dependent on "millions of tons of organic nitrogen fertilizer that were extracted from plant and animal bodies" across the globe, the chemical farming Liebig's theories helped inspire allowed agricultural producers to grow in size and efficiency, expanding cultivation "even in unfamiliar landscapes with very different soil profiles," thus supporting the growth of colonial agriculture in Africa, Asia, and the Americas and allowing capitalism to overcome limits to food production in Europe.[95] As farmers eagerly read and applied Liebig's ideas, their methods helped generate a transnational market for new agricultural chemicals and presaged today's industrialized global food production system.

Liebig's desire to transform eating into a scientific practice undertaken by a mechanistic body tethered to the goal of health and longevity thus gave rise to a geopolitical context in which food was increasingly understood as both a rationale and means for expanding colonial power and the riches it reaped. Concern about how to feed Europe's swelling population and enhance the physical capacities of its labor force centered on the maintenance of healthy soil for growing plants that could be fed to both humans and livestock, but also on people's access to meat. According to historian Wilson Warren, "For nearly all of humanity's existence, meat was not a central component of people's diets," yet that changed markedly during this period.[96] Of course, meat's status atop the western European culinary order was not without precedent, especially among the upper classes, nor was its status reducible to its newly proposed nutritional qualities. Still, the importance of meat in people's diets grew exponentially in the latter half of the nineteenth century, when, according to Warren, "all people in the West, rich and poor, ate more meat . . . than ever before."[97] There were a number of reasons for this shift, including rising incomes, but the invention of protein as a primary nutritional category that was essential to good health provided a scientific justification for meat's cultural allure and the dubious ends to which it was mobilized. Indeed, it was also during this era that meat and protein became synonymous with each other and with vim, vigor, and vitality. While other forms of protein, animal and vegetable, were of interest to Liebig, his research helped solidify an especially tight link between animal flesh, particularly beef, and nitrogenous compounds.

The obstacles to satisfying the growing demand for a year-round supply of fresh, disease-free meat were considerable. They included cattle plagues, rising prices, and the absence of technology that could counter seasonal climates and allow long-distance transportation. It was in these circumstances that pressure to find safe and cost-effective ways to import meat from the European colonies emerged. A physician and prominent commentator of the time, Andrew Wynter, whose work was influenced by Liebig's theories, posed the problem thus:

> What gives rise to the vast majority of disease in our hospitals? What is at the moment deteriorating the lower stratum of the population?— the want of a sufficient supply of nitrogenous food. Those who live by the wear and tear of their muscles are condemned by the present high price of meat to subsist upon food that cannot restore the power that is expended. In the income and expenditure of the human body, in short, they are living upon their capital, and of course sooner or later they must use themselves up. Bread is cheap, because free-trade pours the full sheaves of beautiful foreign lands into our eagerly-spread lap. Why should we not have meat too?[98]

Here the language and logic of political economy that was prevalent at the time is applied directly to human bodies "living upon their capital" in a metabolic ledger of profit and loss, with nitrogenous meat invoked as a potential boon to labor power that, via Europe's colonial reach, was ready in abundance.

As Liebig was helping to stoke the demand for meat, positing the need for affordable, portable forms of animal flesh as "a matter of conscience" for

FIGURE 1.2. Colorful trading cards were used to market and popularize Justus von Liebig's extract of beef, with more than 1,900 different sets (typically consisting of six or twelve images) published between the 1860s and 1970s. The first card, "Scenes in the Life of Liebig," depicts Liebig in his lab at the University of Giessen. In the second card, Liebig's "Real Extract of Meat" is advertised against the backdrop of a "lion hunt in Africa." Public domain.

Western governments, he was also busy experimenting with a solution—an extract of beef that took the form of a thick, dark, potent syrup, similar in flavor to today's Bovril, that could be reconstituted into beverages, soups, and other foods.[99] In the spring of 1862, a German engineer named Georg Christian Giebert, who was building railways in Brazil and Uruguay, wrote to Liebig, whose work he had read, with a business proposition. Giebert described a scene of great waste with the bodies of cattle slaughtered for their hides and tallows but not their meat, which was left to rot in the burning sun.[100] Noting the potential to use this flesh in the commercial production of Liebig's extract, which Liebig was struggling to scale up, Giebert suggested that his correspondent invest in a venture to prepare and sell beef extract on the world market. Other scholars have provided more elaborate histories of the Liebig Extract of Meat Company, but suffice it to say that the founding of the Frey Bentos corporate town on the eastern bank of the Uruguay River, the first global food company according to some commentators, heralded the nascent emergence of a protein supplement market grounded in capitalist and colonial logics of prospecting and speculation.[101] With the ability to process an average of 2,500 cattle a day, and an anticipated production of 1 million pounds of extract by 1868, the factory employed over 1,300 workers, hiring almost exclusively European staff for administrative and technical jobs and local labor for manual work.[102] Although the extract contained only trace amounts of protein—cattle flesh was pressed and boiled into liquid form and its fat removed—it was initially sold as a high-protein supplement, and continued to be associated with the promise of health, repair, and vitality throughout its long commercial life.

The imagined array of uses to which the extract could be put resonates with the malleability ascribed to the protein supplements of today. As Mark Findlay argues in his history of the extract, "Its promoters saw vast potential for this portable food: it allegedly could feed armies in the field, protect sailors from scurvy and meatless diets, enable imperialist explorers to carry the flavour of beef into equatorial Africa, and even increase the productivity of a nation's working classes. Its promoters also believed that beef extract was a medicine. Many physicians and druggists attested that beef teas and beef extracts were valuable in restoring strength to convalescing patients."[103] Indeed, it was in medical circles that the extract initially gained most traction. But as its healthful properties came into question, the emphasis of marketers, and of Liebig himself, shifted to its culinary, flavor-enhancing properties. Made from waste from the leather- and tallow-making processes, the extract also marked the emergence of what was to become a long history of nutritional

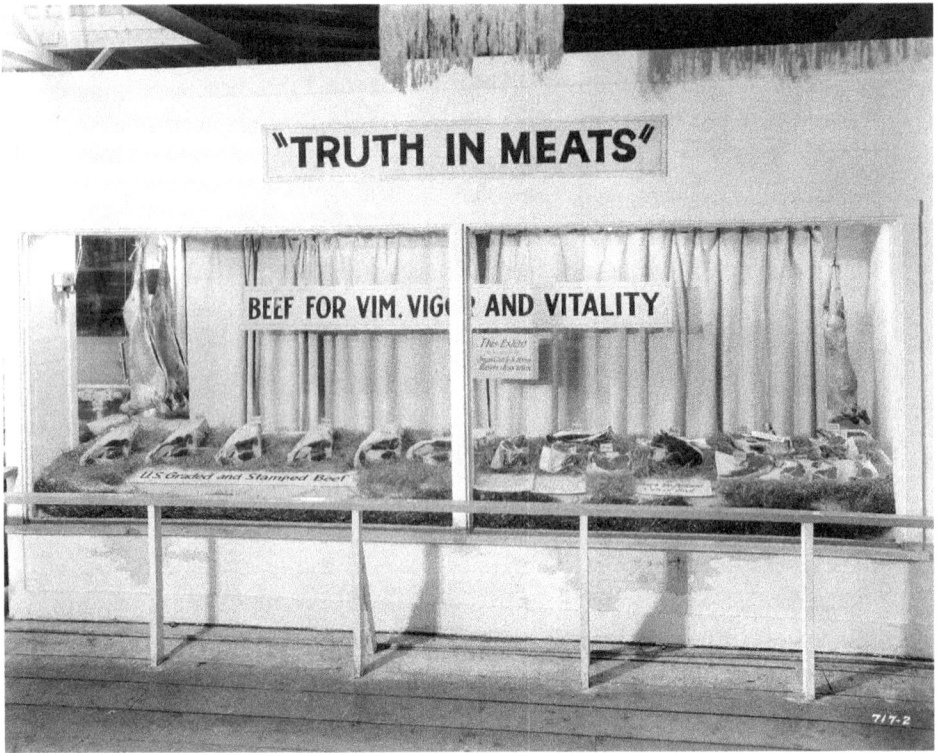

FIGURE 1.3. A livestock exposition display in Oregon during the early twentieth century promoting the health benefits of eating more beef. Extension and Experiment Station Communications Photograph Collection (P120), Oregon State University Special Collections and Archives Research Center, Corvallis, Oregon.

supplements, whereby the excesses and detritus of capitalist overproduction get absorbed into dietary products that are added to, rather than replace, existing consumables, thus growing the market in a sector that struggles to infinitely expand given that people can only eat so much.

Economic justifications for the acceleration of capital accumulation of the sort promoted by Wynter, and commercial enterprises of the sort engineered by Liebig, were complemented by a racial, classed, and gendered science of meat. Edwin Lankester, a surgeon and science popularizer who was influenced by Liebig's work, was typical in his characterization of those who consumed animal flesh as "the most vigorous, most moral, and most intellectual races of man-kind" and thus worthy of their status in the colonial hierarchy.[104] Such discourse was especially strong in the context of British imperialism in India, where the "deficient masculinity" of Bengali Hindu men, in particular,[105] was tied to their vegetarian ways and the distinctive absence of beef in their diet. This view was common not just among colonial officials but also among some Indian nationalists, prominently the Swami Vivekananda, who, according to literary scholar Leela Gandhi, "promoted 'Beef, Biceps and Bhagavad Gita' as the reinforced tonic that would dispel the symptoms of colonial slavery."[106]

Despite his propensity to comment on a wide range of social issues, to our knowledge, Liebig was silent on the ethics of colonialism and the systems of race, slavery, and indenture with which they were entangled. However, his implicit acceptance of a system in which he would become a successful entrepreneur is quite evident in the way he casts about, without any misgivings, for places to procure resources, whether for soil nutrients or meat extract. Moreover, the explicit and exoticizing references to the dietary habits of the "wild tribes" in the *Lancet* series, in which Liebig struggled to explain how nourishment is obtained in places such as China, Japan, and West Africa where "meat is one of the rare enjoyments," work in concert with more subtle allusions to the virility and strength of carnivorous animals—and British and American men—to build a picture that aligns with the broader racial and gender discourse of meat in circulation at the time.

In the decades immediately after Liebig proclaimed that protein was the most significant nutrient, the interest in such science expanded. As Josh Berson explains, the practical goals of research in the late nineteenth and early twentieth centuries were similar to those driving Boussingault and Liebig decades earlier: to improve feed conversion rates in livestock and to understand the energy and especially the protein needs of humans in western Europe. He writes, "The underlying policy concern was clear: How much meat did you need to maintain an industrial labor force?—not to say a modern army and navy."[107] In 1877, German physiologist Carl von Voit, a former student of Liebig, had published dietary recommendations of 118 grams of protein per day for a 70-kilogram man undertaking moderate work, contrary, it bears mentioning, to the findings of his own dietary survey, which suggested that 52 grams per day was sufficient for good health.[108] In turn, Wilbur Atwater, an American chemist who had traveled to Germany to train with Voit, returned to the United States with an enthusiasm for protein that translated into recommendations of 112–50 grams of protein per day—the additional amount reflecting Atwater's belief that American men worked harder than their peers in other parts of the world.[109] Thus, by the turn of the twentieth century, there was general agreement that active men required a minimum of 100–120 grams of protein a day (more than double today's intake when accounting for different average weights across these two eras), primarily from animal sources, alongside an energy intake of 3,000 kilocalories. The suppositions underpinning these recommendations were deeply flawed: Both Voit and Atwater had proceeded under the assumption that if people had the means and intellect to choose their diets freely, they would instinctively opt for more calories. Their "standards," as Carpenter puts it, "reflected what was already being done by the relatively affluent," and, we would add, those deemed racially superior.[110]

The ideology of high protein was not without its detractors. According to Berson, it was challenged most prominently in the West by Yale University physiologist Russell Chittenden, whose 1902 study of student-athletes and army servicemen suggested that much lower amounts of protein (50 to 55 grams a day) and calories could "keep young men in vigor and nitrogen balance indefinitely."[111] Chittenden's work met with a range of responses, favorable and otherwise, but his most vociferous critic was Major D. McCay, a physiologist based in Calcutta who had reported in 1912 on a series of ethically compromised experiments with the diets of Bengali prisoners. To McCay, Chittenden's conclusions were not just incorrect but dangerous, insofar as they undermined the link between an animal protein–rich diet and the "masculine vigor of the more advanced races"; the "evidence of mankind," McCay argued, "points indisputably to a desire for protein up to European standards."[112] Chittenden did his best to refute this logic, noting that his detractors were mistaking cause for effect with regard to the diet of poor Indians: "There are many factors, aside from nitrogen and calories, that play a part in determining proper nutritive conditions," he wrote in a 1911 letter to the *British Medical Journal*.[113] But the idea that it was affluence that led people to eat a protein-rich diet—rather than the other way around— was a hard proposition to sell in this ideological environment, one where the link between colonial vulnerability and "the physical enervation of the malnourished Hindu male body" was firmly established.[114]

Chittenden's perspective on protein was aligned with an adjacent antimeat discourse that Leela Gandhi notes was well established in England by the beginning of the eighteenth century. A thorough accounting of dissident and subaltern approaches to nutrition, protein, and meat is beyond the scope of the present analysis, but our consideration of this context is shaped by Gandhi's study of Mahatma Gandhi's immersion in the English vegetarian movement in the nineteenth century, part of a larger literature revealing how alimentation became what postcolonial theorist Parama Roy calls "an indispensable element" in nineteenth-century assessments of "colonial, anticolonial, and nationalist virtue."[115] This work complicates versions of dietetic history that render colonial bodies as passive and subjugated objects of pathologizing discourses, preferring instead to represent questions of diet as what Roy labels a "terrain for encounter, challenge, transformation, and consolidation" that far exceeded national boundaries, tied as they were to "global and diasporic debates about modernity and its countercultures."[116]

Within Britain, for instance, advocacy for a meatless diet had complex and divergent roots, including a romantic Orientalism that drew on Hindu and Jain traditions of ascetism and nonviolent coexistence with animals, as well as zoophilia, pacificism, and animal welfare advocacy.[117] Both domestically and in the colonies, masculine and muscular versions of vegetarianism existed alongside discourses of feebleness associated with a meat-free diet, and relationships to food across the political spectrum were organized in accordance with class and caste disparities. By the nineteenth and early twentieth centuries, the period about which Gandhi and Roy are writing, vegetarianism was closely tied to revolutionary socialism, anarchy, anticolonial movements, and a vision of interspecies kinship and relationality that directly challenged the scientized, utilitarian, governmental logics that informed the work of Liebig and his colleagues.[118] Thus, while the cultural investment in meat and proteinous diets was pervasive during the latter half of the nineteenth century, it was not the sole preserve of the colonial powers, nor was it without its detractors. Contrary to one unnamed critic's claim that low protein standards were a "mistake" that pitted "Chittenden against the world," challenges to protein orthodoxy from a variety of locales would persist well into the twentieth century, even as Liebig's orientations and interests continued to reverberate.[119]

Conclusion

The preoccupation with protein intake on the part of colonial powers did in fact fade in the early 1900s, before returning in full force toward the middle of the twentieth century and in the form of the notorious Great Protein Fiasco, the subject of the chapter that follows. The acceleration of knowledge about the constitution of proteins in 1902 led researchers to hypothesize that the amount of protein consumed was less important than the right combination of amino acids in concert with the necessary vitamins, a formula that could be met with "ordinary mixed diets" composed largely of whole vegetable foods.[120] Moreover, advances in vitamin research allowed other nutrients to jostle for space with protein in the nutrition science imaginary to an extent that temporarily shifted how the role of protein was perceived. Writing in 1923, prominent American biochemist Lafayette Mendel described the shift as follows: "The recognition of the relative importance of the nitrogenous foods for nutrition ushered in a new era of progress" and "indeed . . . brought about a glorification of the albuminous substance . . . which has persisted in its extreme form almost until the present time. Now, the pendulum of enthusiasm about the proteins has swung from one extreme to another."[121]

Mendel was not wrong about the comparative waning of enthusiasm for protein during this period, but a long historical view suggests that Liebig's persistent preoccupation with protein availability in the face of challenges to capitalist agricultural expansion resurfaces in mid-twentieth-century debates about protein deficiency, population, and the green revolution in the decolonizing world, and again in present-day discussions about eco-capitalist alternatives to industrial meat. Likewise, his interest in attempting to quantify dietary protein requirements remains important to policymakers, health and nutrition researchers, fitness enthusiasts, and dietary adherents today. We have our reasons, then, for having stayed with Liebig, who allows us not only to chart key chapters in the modern history of protein but also to show how the roots of protein's elusive ontological character and adhesive cultural qualities are evident from the formative days of Western biochemistry.

This history reveals that the uncertainties of protein science at a molecular level persist, and as we discussed in the introduction, those uncertainties are mirrored in contemporary debates about how much protein a particular person requires for optimal health and the extent to which it is necessary to focus on protein intake at all if a person has access to enough calories. Yet amid this uncertainty, several of Liebig's fundamental assumptions live on. The idea that a person's protein requirement is based on the amount of physical work they perform; that elevated protein intake is necessary to the formation of productive capitalist subjects and modern, civilized societies; that growth is the primary measure of economic and corporeal "health"; that because we are made of meat, meat or meat analogues provide the optimal form of protein; and that protein is the most significant nutrient persist in one form or another to this day, regardless of what the science says or the degree to which it is contested. While protein remains malleable and adhesive, these are the enduring cultural logics to which it is tightly and persistently tethered. Thus, when Logan Roy, CEO and patriarch in HBO's wildly popular television series *Succession*, inquires about the most significant points to draw from reports and memos by asking, "What's the protein?," viewers know exactly what he means.

2

THE GREAT PROTEIN FIASCO, THEN AND NOW

*Nutrition, Development, and
the Trouble with Growth*

What gives rise to the vast majority of disease in our hospitals? What is at the moment deteriorating the lower stratum of the population?—the want of a sufficient supply of nitrogenous food.—Dr. Andrew Wynter, *Our Social Bees* (1861)

Deficiency of protein in the diet is the most serious and widespread problem in the world. —Marcel Autret, "Encouraging the Use of Protein-Rich Foods" (1962)

Four decades of relative neglect may be ending, as we return to the importance of protein, derived from the Greek adjective πρωτειος (proteios), meaning the first rank or primary. —Richard Semba, "The Rise and Fall of Protein Malnutrition in Global Health" (2016)

In the long history of protein evangelism, one particular chain of events stands out. The Great Protein Fiasco, as it came to be called, references the intense political and economic investment in the idea, subsequently discredited, of a deep and deadly "protein gap" between the world's rich and poor.[1] For Donald McLaren, a medical missionary and trailblazer in the study of nutrition and childhood blindness who coined the phrase in a 1974 article in the *Lancet*, the overemphasis on protein deficiency in efforts to address malnutrition in the Global South had obstructed biomedical progress, led to the underestimation of more pressing nutritional issues such as marasmus (a deficiency of all three macronutrients), and undermined the value of non-Western diets.[2]

McLaren's publication, the climax to a series of articles beginning in the mid-1960s, helped transform approaches to what is now commonly called "severe acute malnutrition," even as protein primacy reemerged in other spaces and at other moments in the decades that followed.[3]

The idea of the protein gap and its subsequent rendering as a fiasco have come to occupy a prominent place in nutrition history. The alleged disparity emerged from a jumble of cursory observations and partial surveys undertaken by colonial representatives across a handful of African countries in the early to mid-twentieth century.[4] Yet the effort to close the gap became a linchpin of international policy and intervention in the 1950s and 1960s. By then, two related assumptions had taken hold: first, that a lack of protein caused a particular form of childhood malnutrition known as kwashiorkor; and second, that protein deficiency was the primary cause of malnutrition globally. As John Nott reveals in his rich history of the politics of nutrition between imperial Britain and Africa, kwashiorkor and protein deficiency were used to explain not only a set of physical symptoms but the problem of "underdevelopment" in the sub-Saharan region and beyond.[5] In this context, providing more protein, especially more protein in forms that aligned with the West's agribusiness interests, emerged as a solution. In effect, protein became a magic bullet that could be targeted at the problems wrought by cultural and economic imperialism, without addressing those underlying systems.

Under the auspices of the colonial administration, and in later years the United Nations, the quest for amelioration propelled several decades of sustained "humanitarian" activity. Attempts to address the "protein deficit" manifested in several ways. These included the construction of a considerable (post)colonial bureaucracy dedicated to measuring and treating kwashiorkor; a reinvigorated discourse of African otherness tied to food and family structure; and two primary practical interventions: the development of a powerful infant formula industry, and the production of a steady stream of highly engineered and doomed-to-fail protein-rich food commodities made from the waste and surplus of the overdeveloping West.[6] The latter, for scholar of humanitarianism Tom Scott-Smith, epitomized the high modernist approach that dominated development in the 1950s through to the mid-1970s.[7] Buoyed by a mindset in which science, planning, and monitoring could provide the optimal diet, nature could be rationalized, and agricultural unpredictability— and societal tradition—overcome by industrial food production, man-made proteins emerged as the key to freedom from hunger. Scott Smith describes the vision this way: "The dream in all cases was to provide a cheap, efficient, mass-produced food that could sustain human bodies in extremis. This was a dream for the space age, which came associated with a vision of life emerging from waste, of human sustenance freed from irrationality."[8]

The dream was not to be realized. Neither the economies nor ecologies of protein in this period were as amenable to rationalization as the nutrient's proponents anticipated. Histories of the fiasco conventionally proffer that after several decades of sustained commitment, but with little progress in addressing hunger and malnourishment, the emphasis on kwashiorkor as the major nutritional problem of the day gave way. In its place, the idea that protein malnutrition primarily emerges as a problem in the context of insufficient calories came to the fore.[9] With this, the focus of internationalist efforts, which were in any case starting to wane as the neoliberal model of development gained ground, shifted elsewhere, and new "charismatic nutrients," such as vitamin A and omega-3, emerged in protein's stead.[10]

In this chapter, we revisit the fiasco with the aim of locating it within the longue durée of protein history. "Fiasco" suggests a temporally and spatially limited debacle, but neither protein fetishization nor its articulation to developmental concerns and colonial logics was entirely new, as the previous chapter makes clear. Nor did efforts to promote protein's value vanish in the decades that followed, as the remainder of the book reveals. Highlighting the persistence of protein supremacy, Kenneth Carpenter writes that "this controversy from the 1950s to the 1960s was really the third of a series of disputes on whether the human diet contained too little or too much protein— the first coming in the mid-nineteenth century and the second at the beginning of this century."[11] Carpenter's painstaking account of the development of protein science, which concludes in the early 1990s, is an important source for the analysis that follows;[12] it is also a useful reminder that while the "history of enthusiasm for protein" is long, it is not absolute or without its fallow periods.[13]

Bringing a concern with the social, contextual valences of nutrition to bear on more conventional histories of the evolution of protein science, this chapter traces the continuities and disruptions in protein's biography with a particular focus on the early twentieth century through to the 1970s. Placing the fiasco in an extended historical frame helps to highlight those aspects of protein's biography that resurface across the multiple sites we explore. Especially crucial in this sense is protein's status as a solution to underdevelopment of all kinds and its attendant association with growth—growth of infant bodies into healthy adult ones, of soft tissue into hard muscle, of small enterprises into global markets, and of "backward" and peripheral colonial economies into "modern" engines of transnational capitalism.

Protein has featured in discussions about "development" (itself a multi-plicitous and politically laden signifier) since it was first identified as a growth accelerant by protein evangelist Justus von Liebig in the mid-nineteenth century (see chapter 1). Aya Hirata Kimura points to this history in order to make the case that the long-standing "special status" of protein in Western culture "perhaps accounted for the initial demarcation of it as the signifier of what was missing from the non-West's diet" when protein emerged as the symbol of the "Third World food problem" in the mid-twentieth century.[14] Here Kimura draws on the work of Geoffrey Cannon, who emphasizes the colonial inflections of Liebig's research, arguing that the value Liebig placed on growth—of plants, farm animals, humans, societies, and economies—derived from a philosophy that viewed the living world as a resource that can be "conquered and controlled."[15] In the heyday of industrial capitalism, growth "in every sense" was understood as a "self-evident" good and imperial governments seeking the production of more plentiful, bigger, and faster-growing crops, animals, and humans found in nutrition science the justification, and in some cases the means, by which to achieve such expansion.[16] As Cannon writes, "Once protein was isolated and identified as the primary or master nutrient and so the nutritional expression of the dominant European ideology, food systems engineered to emphasize animal protein had the power to change the world, as they have done."[17]

From its inception, then, protein was inextricably bound up with the colonial project and the mutually imbricated expansion of European bodies and territories it implied. In the early decades of the twentieth century, the preoccupation with growth and size was extended beyond the metropole. During this period, advances in nutrition science, shifting labor needs, and concerns about the insufficient incorporation of colonized regions of the world into the machinery of capitalism (aka "development"), pulled the role of protein in the bodily optimization of colonial subjects to center stage. In keeping with our overall analysis, we focus on protein's adhesions across two key periods—1900–1949 and 1949–75—showing how gendered and racial thinking about nursing mothers and the ideal height of infants helped propel the idea of a protein gap. Our argument is not that protein deficiency or unequal access to it was a fiction; rather, protein supplementation got mobilized as a universal, market-friendly, technocratic solution to a set of contingent colonial, social, agricultural, and labor conditions that underpinned malnourishment. The extent to which protein deficiency was a driver of poor health in this period is hard to ascertain and remains contested even in contemporary research and policy in global nutrition. Our goal is not to adjudicate much less resolve this debate; it is to show how a combination of forces, including growing investments in nutrition research and food technology, an emerging international public health regime, and Western agricultural overproduction, together conjured protein as a solution, in ways that reproduced globally stratified dichotomies between healthy and unhealthy bodies.

Our focus is on the development of Western, (post)colonial governmental research, policy and practice, rather than on the ways in which the idea of a protein gap played out "on the ground," or how protein interventions were understood, embraced, or resisted by local communities. Of course, these realms of social life are inseparable in practice. The fact that an early focus on cow's milk and a later focus on high-tech foods failed as solutions to protein deficiency is a testament to rejection and refusal by those whose diets were deemed lacking, and to the persistence of critics of the protein gap theory, most of whom were from the Global South. The failure to scale up protein enterprises also played a role, though we point readers interested in that part of the story to extant accounts.[18]

A rich body of historical work trained on specific geographic locales during particular periods speaks in more depth to the point of view, activities, and agency of Indigenous Africans, to local nutritional knowledge and concessions made to it by colonial authorities, and to the ways that communities adapted to or subverted attempts to harness their labor through food and agricultural interventions.[19] While none to our knowledge focus on the research and programs that comprised the protein fiasco, this work generates an important counter-archive in which the communities subject to colonial research and interventions come to the fore.

We remain alert to these tensions and currents in what follows. But in keeping with our interest in "studying up," we attend to the long arc of Western history and systems of power, contending that protein evangelism in the postwar era helped keep its subject at the center of nutrition discourse well beyond this period. We also argue that protein persists as a norm against which all other dietetic categories are measured, even at moments when its supremacy is most intensely contested or understood to have been irreparably undermined. "Fiasco" implies exactly such irrevocability—a failed idea never to be repeated—but protein-centrism is alive and well, tethered to an industry of a size and scale that far exceeds its supposed zenith fifty years ago.

Racializing Protein, Pathologizing Mothers: The
Idea of the Gap Takes Hold, 1900–1949

The pattern of logical leaps and erroneous assumptions by which the idea of a protein gap first took hold was underpinned from the outset by colonial thought and practice. The early decades of the twentieth century had been marked by a series of crucial developments in nutritional science and accompanied by great optimism about the implications of emerging knowledge for addressing ill health on a population scale.[20] The idea that major diseases could be eliminated by relatively small adjustments to the diet, or that there might be a set of minimum requirements for keeping those faced with starvation alive, held considerable social and political promise. The identification and synthesis of a number of key vitamins and minerals was especially important in transforming nutritional medicine into "an overtly political concern" and vitamins into "biopolitical objects" by which states could address problems of food supply and public health.[21] In the nineteenth and early twentieth centuries, hunger and its consequences had been primarily viewed as a European problem by European powers seeking to ensure an adequate supply of healthy labor at home and to maintain the military might that underpinned colonial rule overseas.[22] But this started to change with what Michael Worboys calls the "discovery of colonial malnutrition" in the interwar period, when the idea that dietary problems could be fixed by the application of scientific principles was expanded beyond the metropole.[23] As Nott writes, "Extending a politicized concept of nutrition beyond British borders was readily accepted in the context of empire. Investigations into nutrition in European possessions overseas found favour amongst Whitehall politicians already primed to understand the physical capital of the colonised as an extension of colonial power."[24]

In the early days of research in the colonies, the emphasis was primarily on questions of labor power and the value that could be extracted from a well-fed workforce. Nott quotes William Ormsby-Gore, the undersecretary of state for the colonies, who, during a 1926 visit to the Gold Coast, declared that "the capacity of labor . . . is bound up with the question of food. There are few parts of the world where the study of dietetics is more important than in Africa."[25] Since Liebig had first preached the "gospel of protein" in the mid-nineteenth century,[26] the nutritional practices of the relatively large white European male manual workers who formed the subjects of most foundational studies had been understood as ideal and universally applicable.

These vigorous European subjects haunt the text of one particularly influential but indelibly flawed 1927 study driven by a concern with the fitness of "native workers."[27] Coordinated by two British scientists, John Gilks, head of the Kenyan Medical Service, and John Boyd Orr, a Nobel Prize–winning nutritional physiologist who was later to become the inaugural director-general of the Food and Agriculture Organization of the United Nations (FAO), this study was the first attempt by the British to address nutrition as distinct from hunger or famine in the colonies.[28] Drawing on findings gathered by colleagues on the ground, Gilks and Orr compared the impact of diet on the "physique and general health" of the Kikuyu and Maasai peoples. Worried about the labor productivity of the Kikuyu, who food researchers Tamsin Blaxter and Tara Garnett claim were suffering from persistent health problems linked to ulcers associated with poverty in tropical regions, Gilks and Orr attributed the supposed strength and robustness of the Maasai to the prominence of meat, milk, and blood in their diet, and the "unsatisfactory condition" of the Kikuyu to their largely cereal diet.[29] Historian Cynthia Brantley highlights the impact of Orr and Gilks's research, noting its centrality to the first major League of Nations report on nutrition and public health, in 1935, for which it "provided the main evidence for the interpretation that all 'colonial populations,' most of whom ate vegetarian diets, in general were undernourished."[30] As Brantley reveals, this conclusion was unsupported (even by Orr and Gilks's own findings), and allowed to emerge only because positive findings about grain-centric diets were buried and thus ignored. In overlooking the "vast variation" among the diets of men and women and "important distinctions" among adults and children, while exaggerating and oversimplifying supposed differences between "tribes" and between pastoral and agricultural communities, Brantley claims the study gave "scientific authority to the high value placed on animal protein" and "distorted the direction for African nutritional development."[31]

The distorted framework emerged from research saturated with sweeping and flawed assumptions about the social organization and gender norms of "tribal" societies and the character of "African" dietary practices.[32] The impact of land theft by European settlers—a crucial factor in shifting diets—was also left unexamined in Orr and Gilks's study. They incorrectly assumed that they were observing abiding dietary patterns undisrupted by colonization, and noted hopefully that land appropriation might serve to "hasten the improvement of the physical condition of the native and to increase his importance as an economic factor."[33] Within this context, it is unsurprising that an increased supply of the animal foodstuffs that were so valued in the diets of the colonial powers emerged as a commonsense solution to the problem of malnourishment on the African continent. For Orr and Gilks, milk, which was understood as not just a reliable source of calcium but also a "first class protein," was the preferred antidote—a conclusion undoubtedly shaped by the financial support Orr received from the Empire Marketing Board, which was seeking ways to dispose of surplus milk, the gravest agricultural problem facing Britain at that time.[34] In other missives from British officials stationed in Africa, meat emerged as the solution, "especially for those engaged in hard physical labor."[35]

Milk and meat persisted as idealized dietary fixes over the decades to come, but with the unraveling of formal empire in the post–World War II period the emphasis of colonial nutrition shifted away from questions of labor toward questions of "development." With this, the notion that inadequate nutrition and not just infectious disease stood in the way of full participation in the system of global capitalism came to the fore.[36] It was in this context that Cicely Williams, a white, Jamaican, Oxford-educated pediatrician stationed in southern Ghana, provided the first officially documented, English-language description of what Ga-speaking Africans called "kwashiorkor"—a deficiency disease marked by peripheral edema, wasting, and diarrhea.[37] From 1929 to 1936, Williams was employed as a "Woman Medical Officer" in what the British called the Gold Coast.[38] Charged with treating acutely ill infants and children, Williams had noticed a spike in sickness and mortality among two- to four-year-olds immediately after a younger sibling was born and the older child was switched from breastfeeding to the conventional diet of a porridge made with maize or cassava. Among the symptoms these children exhibited were protruding lungs, a distended belly, swollen feet, decolorization of the skin, and a reddish tint to their hair. According to Williams, who published her initial observations in the *Archives of Disease in Childhood* in 1933, the children were receiving an adequate amount of food but were nonetheless suffering from what appeared to be a disease of malnutrition.[39] Following a detailed description of the affected children and their diets, Williams wrote only that "breast milk is probably deficient in some factors, which are at present uncertain. As maize was the only source of the supplementary food, some amino acid or protein deficiency cannot be excluded."[40] Yet from this extremely tentative suggestion, the idea of a relationship between symptoms of dietary deficiency and lack of protein began to take hold.

Initially, the impact of Williams's missives was muted. By Carpenter's account, each of Williams's articles prompted a patronizing rejoinder from H. S. Stannus, an authority on deficiency diseases in the colonies who disagreed with Williams's conclusions and disliked her use of an Indigenous descriptor (he thought she was seeing pellagra).[41] A 1935 research note from another medical officer agreeing with Williams about the unique character of the symptoms helped prolong the conversation.[42] So did at least eleven papers from different parts of the world, describing cases with similar characteristics to kwashiorkor, which emerged in the two years following the publication of Williams's first paper.[43] During this time, debates continued about whether the purported symptoms represented a form of pellagra or a new disease and what the best term was to describe them. While it took another two decades for scientists and policymakers to build on Williams's research in earnest and for a more solid consensus to emerge, Nott argues that Williams's work provided a crucial prompt: "From this point until past the end of empire, kwashiorkor dominated research into and discourses around food and health in Africa, its apparent prevalence confirming a continent-wide 'protein-deficit,' the widespread need for protein supplementation and contributing, ironically, to the subsequent increase in clinical deficiency."[44]

There are a number of possible explanations for the influence of Williams's quite tentative suggestions. While the newness of kwashiorkor in the annals of colonial medicine, if not in Indigenous Ghanaian knowledge systems, surely fed growing interest in the condition, the longer history of amino acid ascendance helped lay the foundations well before Williams's "discovery." The work of Liebig in establishing protein, especially meat, as the most important nutrient continued to hold sway in the face of occasional challenges from researchers who warned of the dangers of high-protein diets. But Nott explains that Williams's work also caught her readers' imaginations because of the lengthy and exoticizing descriptions of the symptomology of kwashiorkor (a term she didn't use until her second publication on the subject, in the *Lancet*, in 1935).[45] Her focus on "depigmentation" and her intricate descriptions of ideal black skin as opposed to what she observed about the tone and texture of the skin of her patients aligned with the "established otherness" of African bodies and lives, tropes that Nott claims were "only exacerbated by Williams's use of and, subsequently, the general adoption of the Ga nomenclature."[46] Both Williams's papers, but especially the first, also included objectivizing and dehumanizing photographs of sick and dead children that, while typical of medical communication at the time, fed into the exoticization and exceptionalism that characterized the era's tropical medicine.[47] As Kimura writes, "With its striking visual signs . . . kwashiorkor powerfully symbolized the misery and poverty of the Third World, which further cemented experts' fascination with protein."[48]

As the hosts of the podcast *Maintenance Phase* make clear in their critical retelling of the Great Protein Fiasco, there is no sugarcoating Cicely Williams's racism and investment in the colonial project, even if this is buried in the vast majority of accounts of her work, which tend to take a liberal feminist perspective that highlights her "enlightened but paternalistic" approach.[49] Williams's musings on the lack of initiative, honesty, and imagination of "the African," which she noted at one point "may be ascribed in part to some dietetic imbalance and some to imbalance of upbringing," are usually omitted from accounts that celebrate her socially grounded approach to pediatric work.[50] The latter was notable, Nott explains, because she "listened to her patients," especially the mothers and grandmothers who apparently brought kwashiorkor to her attention and likely inspired her decision to use the Ga term for the condition. Nott notes that Williams engaged with the people she encountered "with a degree of sensitivity uncommon in colonial physicians."[51] This is why Ghanaian medical leaders have spoken favorably of Williams in their attempts to counter the exoticism, error, and nutritionism that have characterized medical discourse on kwashiorkor. As prominent Krobo/Dangme/Ga physician and scientist Felix D. Konotey-Ahulu wrote in 1994, "Nobody in more than half a century had . . . improved upon her work."[52] Writing again in 2005, he said, "Clinical epidemiology, i.e., answering the questions Who? Which? Where? When? Why? What? And How? is far and away the best tool to investigate a tropical phenomenon such as kwashiorkor, and Cicely Williams gets my full marks for employing that tool with hardly any funds for medical research."[53] The same inference might be gleaned from reviewing the current state of knowledge on kwashiorkor today. The condition remains "ill defined," according to the *British Medical Journal*, with researchers continuing to debate the extent to which protein deficiency, recent weaning, and a maize-based diet—all factors that Williams pondered—underpin symptoms.[54]

Williams's influence on international development politics did not end with her tentative observations about kwashiorkor, and in fact, the other issue enduringly linked to her legacy—the politics of infant feeding—was itself inextricably bound up with the politics of protein. Williams left Ghana for Malaysia in 1936, following an argument with her boss about the treatment of a child with tuberculosis, and soon became concerned about the high rates of infection and malnutrition she observed among infants there. Williams had been an early proponent of bottle feeding, even arguing to her bosses in London that she should be permitted to advertise Nestlé tinned milk in her clinic in Accra. But upon learning that many Malaysian mothers were giving their children sweetened condensed milk sold to them by women employed by Nestlé to dress up as nurses and spread the message that their product was superior to breast milk, Williams became a crusader against breast milk substitutes.[55] This excerpt from a 1939 talk in Singapore titled "Milk and Murder" reveals the depth of her disdain: "If your lives were as embittered as mine by seeing day after day this massacre of the innocents by unsuitable feeding, then I believe that you would feel as I do, that misguided propaganda on infant feeding should be punished as the most criminal form of sedition and that those deaths should be regarded as murder."[56] While a deep engagement with the complex (anti)colonial politics of lactation and bottled milk is beyond the scope of this chapter, the intimate link between this issue and the emergence of the protein fiasco is worth highlighting. In Nott's words, "Protein, and its apparent absence, lies at the heart of this enduring history," a claim that podcast host Michael Hobbes elaborates as follows:

> This idea of baby formula instead of breastfeeding is intimately wrapped up with . . . protein deficiency because the entire idea is that people are getting enough calories, but they're not getting enough protein. That is essential to the reason why these formulas are being pushed on mothers. . . . This entire understanding of protein is that if you eat the local food, your baby is going to get kwashiorkor. So it's really important that you buy this Nestlé stuff instead of the food that you already have access to that doesn't cost sixty bucks.[57]

The eventual diagnosis of kwashiorkor as a form of protein deficiency that existed on a world scale thus opened the door to a global supplementation industry offering a quick fix to a problem framed as an outcome of uneducated mothers and unhealthy diets. For a variety of reasons, the fix was never realized, the most crucial determinant being a price point that forced caregivers to stretch minimum servings over several days thus leaving the infant undernourished. But posing culturally induced protein deficiency as the problem, and formula as the solution, allowed colonial states in an era of transition to circumvent responsibility for malnutrition while reinforcing gender and racial hierarchies that underpinned their waning authority.[58]

We will return to Williams's legacy shortly, but it is important to emphasize that the relative veracity of her ideas seems to have had little to do with how they got taken up in the years that followed. More crucial was the fact that Williams's reports appeared at a moment when the idea was taking hold that "the native food problem" was "not so much one of quantity as quality."[59] This assertion was articulated in an influential 1939 report, *Nutrition in* the *British Colonial Empire*, and bolstered the singular focus on kwashiorkor, which had been defined as a problem of protein deficiency, and thus on the nutritional makeup of the individual diet—an issue that the imperial power thought could be fixed with relatively minimal intervention.[60] In contrast, undernutrition, from which colonial administrations shied away as an explanatory framework, represented a more systemic problem that could not be addressed within a reductive nutritionist paradigm.[61] Framing kwashiorkor as the result of "native" ignorance and backwardness rather than an economic system built on low wages, resource extraction, the disruption of local food systems through the implementation of cash cropping, and the imposition of vast inequalities in wealth and status, allowed the imperial regime to evade responsibility for any increases in hunger, which were frequent in the years that followed.

Speaking to the political exigency of a focus on dietary composition, Nott writes that the preliminary reports that informed *Nutrition in the British Colonial Empire* were revised to remove any hint that the colonial system was driving pervasive malnutrition.[62] This included missives from the Gold Coast that suggested, contrary to subsequent publications, that the most pressing nutritional problem was not a lack of quality foods but severe food shortages and the ongoing threat of famine.[63] While the Ga understanding of kwashiorkor as the "disease the deposed baby gets when the next one is born" was to be gradually squeezed out of the narrow medical framing that took hold in the decades to come,[64] Nott explains that a 1954 study listed thirty-two names for similar collections of symptoms across several African countries, "most of which suggest the same etiology of the disease—short birth spacing and short breastfeeding duration."[65] In other words, while it is possible that protein deficiency was a problem in some places during this period, there is a strong case to be made—as Nott demonstrates—that "the gendered pressures of colonial government," which included pronatalist policies discouraging protracted breastfeeding and sexual abstinence, underpinned what Williams observed in her local practice.[66]

The reduction of the syndrome to a matter of poor diet, with little heed paid to the role of colonial policies in producing malnutrition, had several lasting effects. Symbolically, it allowed for the depiction of kwashiorkor as a failure on the part of African mothers, communities, and cultures by devaluing local knowledge and childrearing practices and pathologizing local diets. Nott refers to the recurring trope of an inadequate "African diet" linked to the "backwardness" of the African people, even as discussions of kwashiorkor were increasingly couched in medical language.[67] Among the growing community of researchers studying the disease in the late 1940s and 1950s, Hubert Carey Trowell, a British physician stationed in Kenya and Uganda, was especially prodigious, and his writings exemplified this ambiguity. Trowell acknowledged that the syndrome (a label he preferred because kwashiorkor "is probably not a clinical entity") was "recognized by many African tribes," that it was related to the timing and environment in which weaning occurs, and that it was compounded by social conditions such as poverty and water shortages.[68] But he also returned repeatedly to the "faulty" African diet and to ill-informed choices on the part of African mothers. To Trowell and his coauthor, Jack Davies, a pathologist who became a leading authority on kwashiorkor, a faulty diet comprised "little protein and fat, and much carbohydrate and roughage," with one primary obstacle to an adequate diet being the absence of cow's milk.[69] Drs. John Fleming Brock and Marcel Autret, the authors of a landmark 1952 World Health Organization (WHO) report, were similarly preoccupied with the curative potential of animal milks; they also focused on what was perceived as miseducation, claiming that "with the best intentions, African mothers commit many grave faults in the feeding of their children."[70] Brock and Autret went so far as to propose that "it would not be too far-fetched to attribute protein deficiency, at least in part, to the backwardness of the African people."[71] According to Nott, "even in 1979," Davies asked a Johannesburg audience, "How much did this nutritionally-induced apathy contribute to the docility of Negro slaves?"[72] Aubrey Gordon argues that it was with this mindset and in this environment that the idea that "we have to get protein to poor people" took hold.[73] Reducing kwashiorkor to a problem of nutrition, and placing a racialized notion of protein deficiency at the center of development efforts, also provoked what was to become a primary and highly contentious "solution" to the supposed protein gap: a transnational push to develop synthetic, high-protein foods that could supplement or replace local diets.

Protein as Savior: Food Surpluses, Nutritional Supplements, and the Rise of an International Public Health Regime, 1949–1974

The scale-up of such responses was made possible by the rapid development of international organizations in the wake of World War II, especially the FAO in 1945 and the WHO in 1948. At the 1949 inaugural joint meeting of the agencies to coordinate their nutritional plans, kwashiorkor was the second item on the agenda. Records suggest that officials were concerned about "disturbingly high" mortality rates in some parts of Africa and that they were starting to postulate that kwashiorkor in infancy was to blame for high rates of cirrhosis among adults in Africa and Central America.[74] While Carpenter notes that there were "no suggestions as to what nutritional deficiencies might be responsible for the condition," the joint committee nonetheless "urged that more animal milk should be made available for weanlings." They went on: "Where that was difficult, the production and use of foods and/or preparations which can act as a partial substitute for milk should be vigorously encouraged.'"[75] This was a period in which the United States was amassing a huge surplus of milk and struggling to find ways to dispose of it (see chapter 3). As protein critic Donald McLaren wrote, "It was clearly more satisfactory in every respect to dump it in developing countries than to have to bury it in the United States as was contemplated by the Department of Agriculture at one point."[76] Deemed a cheap and efficacious protein for "making good the deficiencies in the diet of the poor," millions of tons of dried cow's milk were shipped to the Global South over the ensuing decades.[77]

While protein-as-milk was lurking in the background as a solution in search of a problem throughout this period, its emergence as a dietary champion was not straightforward. Propelled by a resolution at the 1949 meeting recommending that the WHO conduct an inquiry into kwashiorkor, South African physician Brock, and pharmacist Autret (leader of the nutrition division of the FAO from 1960 to 1971), undertook a tour of Africa. Brock and Autret's report is thorough, and at moments tentative and nuanced. For instance, at one point they write, "Knowledge of the nutritive value of African foods is steadily increasing but still remains insufficient. In particular, little is as yet known of the amino-acid content of the proteins in the vegetable and even animal foods which are included in African diets. . . . In the existing state of knowledge, it is hard to establish a precise relation, in scientific terms, between kwashiorkor and nutritional factors."[78] They also note tensions around cash cropping. Brock and Autret offer no criticisms of colonial economic policy, and ignore entirely the destruction of Indigenous food systems and correlated health impacts wrought by the shift to export economies, but they do include in their discussion of kwashiorkor prevention "the importance of increasing the production of food crops for local consumption" and note that "the policy adopted should take account of the real needs and interests of the African in each individual territory."[79] However, the apparent absence of kwashiorkor in areas where there was a relatively plentiful supply of meat, fish, and milk, and the apparent success of treating symptoms with skimmed milk or "good-quality vegetable protein sources," led Brock and Autret to identify protein deficiency as the cause of the condition.[80] Their report concluded with what was to become a much-quoted statement: that kwashiorkor is "the most serious and widespread nutritional disorder known to medical and nutritional science."[81]

The authors reached this conclusion based on visits to just ten hugely diverse African countries and, as McLaren pointed out in the exposé that gave the protein fiasco its name, made few references to other deficiency diseases or to different parts of the world, despite the universal nature of their conclusion.[82] Brock and Autret's report was accepted at the second session of the joint committee in 1951 and resulted in a commitment to increasing the provision of locally produced plant proteins to combat the syndrome. A third session was held the following year, where reports of the condition from "an increasing number of countries," including India, were discussed. Concomitantly, a more singular focus on protein emerged along with the use of the label *protein malnutrition*.

A conference on protein malnutrition, held in Jamaica in 1953, drew twenty-six participants "from every continent" as the idea of a crisis spread.[83] Although the beat of the protein drum persisted, not everyone present was convinced that protein malnutrition was responsible for the common set of symptoms under discussion, and participants acknowledged that there were local variations in symptomology that were deemed to be "caused by deficiencies in one or more of half a dozen nutrients."[84] Prominent Indian nutritionist Coluthur Gopalan was among the skeptics, noting, as he continued to do throughout his long career, that the diet of children growing up in poverty tended to be low in calories. In taking this position, he and his allies ensured that a lack of calories persisted as a possible explanation, along with the role of infection, as a stimulant to kwashiorkor's emergence.

According to Carpenter, one point of consensus centered on the idea that skim milk had proven to be the most effective treatment "everywhere," though attendees agreed that locally available plant sources would need to be developed, and they learned of research in both Mexico and India showing that bean-based mixes helped cure kwashiorkor symptoms. But with a surplus the United States was eager to unload, milk persisted as the preferred solution. To this end, delegates learned that during 1953, a by then seven-year-old United Nations International Children's Emergency Fund (UNICEF) had distributed enough dried skim milk to provide approximately 9 grams of protein per child daily to Belgian Congo and Ruanda-Urundi.[85] In this instance, "it was judged that the condition of the children had improved."[86] UNICEF's role in milk promotion predated international concern about protein, but

FIGURE 2.1. A shipment of powdered milk, funded by the Canadian government, arrived in Bombay on October 15, 1959, as part of the UNICEF milk feeding programs in India. Workers unstack bags of powdered milk aboard the vessel. UN Photo.

the perception of the protein gap and responses to it aligned neatly with the ideology that "civilization followed the cow."[87] Scott-Smith writes, "In its earliest days, UNICEF had, in the words of its official historian, been a 'giant organizational udder': a proselytizer for milk, a worshipper of milk, which was presented as the key to nutrition."[88] While UNICEF remained focused on milk as a category well into the 1960s, it gradually began to incorporate the idea of milk substitutes into its efforts as the narrowing of a focus from food to nutrients, and to protein specifically, intensified.[89]

The rush of protein-focused meetings continued with gatherings in 1954, 1955, 1957, and 1960. While members at the 1955 conference agreed to lower the recommended protein allowance for adults, claiming that previous levels based on intakes in Western countries were unrealistic and had no scientific basis, contestations remained anchored within the nutrition science paradigm.

As a result, considerations of the social and economic conditions that might be producing malnourishment were largely excluded from the agenda. It was at the 1955 meeting that milk was proposed as a "reference protein" to determine the amino acid requirements for infants and young children.[90] It was also here that Autret noted that the FAO's first priority was to develop milk production wherever that was practicable. This priority reflected the aspiration of partners such as UNICEF to move away from emergency feeding using surplus European and US cow's milk and toward the establishment of a dairy industry in the Global South.[91] But Autret was well aware of how difficult that would be, and went on to say, "The next approach is how to use for feeding children various protein-rich foods now used for animal feeding only. Of one million tons of fishmeal now being produced in the world each year, 30 to 40 per cent . . . could be used for feeding children. In Africa . . . 500,000 tons of peanut meal are available, which might well be turned into an edible protein food; there are also large quantities of soy flour."[92] It is here that we see a focus on culturally alien, techno-foods "from nowhere" emerge as solutions to the alleged protein gap.[93] Gopalan pushed back with a systemic critique, arguing that "a greater contribution to the prevention of protein malnutrition in certain regions of the world may lie, not in supplying supplements, but just in eradication of malaria, the improvement of poor home economic conditions, etc." [94] But the march toward new protein-rich supplements continued apace. Some experts voiced concerns about the safety of untested materials. But according to Carpenter, others argued that "where similar products had been fed for years to young chicks and pigs, it seemed unreasonable to hold them back from infants."[95] Thus, when the joint committee met for the fifth time in 1957, it announced that it had now moved from the research and analysis stage to the development of preventative measures "including the effective use of protein-rich foods other than milk."[96]

The shift toward the development of new commodities had been further consolidated in 1955, when the WHO established the Protein Advisory Group (PAG) of the United Nations. With financial support from the Rockefeller Foundation and UNICEF, PAG was charged with advising the FAO and UNICEF on the "safety and suitability of proposed new protein-rich food preparations" to combat undernutrition.[97] The criteria for the development of these preparations reflected the principle that locally produced, culturally acceptable, safe products that could supplement rather than replace existing foods were the desired outcome. However, this goal proved to be hard, if not impossible, to fulfill, despite efforts to enlist appropriate expertise and the investment of considerable resources. Clinical nutritionists on PAG were soon joined by food scientists, technologists, and marketing experts. Social scientists were later invited to join the effort, with a 1960 conference in Cuernavaca, Mexico, emphasizing the importance of understanding the culturally specific character of diet when introducing alien foods.[98] The arrival of social considerations did not translate into bigger questions about the underlying premise of the enterprise, however. For it was also at this conference that leading Mexican pediatrician Federico Gómez Santos proposed that "growth failure, based on weight-for-age, be used to define the prevalence of protein-calorie malnutrition."[99] In back-to-back conferences in 1963 and 1964, a sense of urgency and a focus on infant size continued to build, and the report from the 1963 conference proclaimed that "the present situation was so grave that it should be recognized as an emergency."[100]

By 1968, international organizations had declared a global emergency. A UN report titled *International Action to Avert the Impending Protein Crisis* made the argument that an inadequate protein supply was a threat not just to public health but also to "world peace and stability."[101] Two years later, the FAO framed the problem similarly in a report titled *Lives in Peril: Protein and the Child* (1970).[102] By this time, it was evident that US milk surpluses were much smaller than predicted and that generous supplies of powdered milk had begun to evaporate.[103] Both reports thus emphasized supplementation, with recommendations that included fortifying ordinary sources of plant protein with essential amino acids (the UN report) and promoting protein food mixes (the FAO report). In 1971, a UN panel of experts followed up with *Strategy Statement on Action to Avert the Protein Crisis in the Developing Countries.*[104] These documents included claims that in some areas up to 50 percent of children aged one to fifteen years were suffering from protein-calorie malnutrition, with "stunted physical growth" the symptom that most often emerged as the focus of concern. This panel built on the 1957 proposals by adding a focus on genetically engineered single-cell protein sources, synthetic amino acids, and the development of research and training centers and initiatives to deal with the crisis. The estimated cost of implementing these proposals, according to Carpenter, was $75 million, "but they were well received by the General Assembly and the UN Secretary General," and the work of the PAG was broadened: The World Bank became an additional sponsor in 1971.

The declaration of a global emergency opened the floodgates to Western governments and their industry partners eager to build a market for high-protein foods. While a 1966 report from the FAO/WHO Expert Committee on Nutrition lamented that industrial production of protein concentrates was "slow and uncertain," a steady stream of unappetizing prototypes was developed during this period.[105] These included a single-cell protein cultivated on oil, developed by British Petroleum; a leaf-protein concentrate generated by putting inedible leaves through a centrifuge; a fish-protein concentrate made from offal left over from filleting; and Chlorella, an algae grown on sewage.[106] Noting the particular unpopularity of Chlorella, Nott writes, "Nonetheless, as food grown entirely on waste, it did represent the crowning achievement of modernist nutrition and is still available, marketed as a health food."[107] These concoctions all ultimately flopped. But they nonetheless helped consolidate what Tom Scott-Smith calls the "high-modernist" approach to nutrition, in which culturally alien, high-protein foodstuffs intensified the "continuing dislocation of nutrition from food" and undermined communities whose ability to feed themselves had been decimated by the enforcement of colonial trade priorities.[108]

Protein's Fall, 1974 and Beyond

Efforts to address the protein gap through techno-supplements may have floundered, but the idea of the gap, and its urgency, continued to build into the early 1970s, even as long-standing pockets of resistance to the assumption that protein deficiency was the main driver of malnutrition began to take hold. Skeptics included Cicely Williams herself, who had spent years trying to "debunk kwashiorkor," or more specifically, the idea that the disease emerged from a singular deficiency of protein.[109] As Konotey-Ahulu wrote of a 1986 meeting with Williams, "We were both greatly baffled why so many of today's experts (especially those who have little experience of a tropical sojourn) find it difficult to accept that 'kwashiorkor is the result of a social pathology before it is the outcome of a biochemical pathology.'"[110] Konotey-Ahulu maintained that kwashiorkor was an outcome of the "birth position" of the sufferer and could be fixed if parents were able to afford a more varied and proteinous diet for children who were still being weaned when their next sibling was born. For him, Williams, and indeed for many of today's experts, the problem wasn't and isn't just amino acid deficiency, but some combination of birth order, financial means, a maize-only diet, and lack of protein. In this view, kwashiorkor is a highly contextual syndrome (as most are). But the social and regional specificity of the diagnosis got lost when it was scaled up into international guidelines that failed to reflect the evidence that was available. These guidelines were then applied not just to the rest of Africa—in itself an unfathomable leap—but to the developing world as a whole.

Konotey-Ahulu's skepticism gives weight to Nott's claim that doubts about the causes of kwashiorkor and the idea of a protein gap emerged largely from those living and working in the Global South, doubts that in the short term continued to be overridden by the "Western medical consensus."[111] For example, C. Sathyamala, a physician-epidemiologist and development studies scholar, explains that framing kwashiorkor as a global problem was a spur for Indian nutritionists such as Gopalan and P. V. Sukhatme, a statistician with the FAO, to put forth a counter-discourse questioning the generalization.[112] Gopalan's early challenge to the understanding of kwashiorkor as a purely protein-deficient state later came to be bolstered statistically by Sukhatme, who found that a cereal-based diet adequate in calories was also adequate in protein. In a retrospective article on his role in ending the fiasco, McLaren acknowledged that his "was by no means a single-handed protest" and also pointed to the work of Sukhatme, who had been insisting during the same period that "food-consumption data pointed to a deficit of energy rather than of protein."[113] Alluding to the ongoing influence of former colonial officers even after colonial governments had been overturned, Ghanaian medical pioneer Fred Sai noted with reference to Ghana's nutrition program, "You can overthrow a government very much more readily than you can change members of the civil service."[114]

The early 1970s was also a period when internationalist, leftist campaigns focused on global inequality and the systematic underdevelopment of the Third World by colonial and neocolonial powers gained ground. US pediatrician Derrick Jelliffe's 1972 work on "commerciogenic malnutrition," a term he used to describe malnourishment resulting from the "thoughtless promotion" of commercial milks and infant foods, was especially influential.[115] While hunger had long been recognized as a market to exploit, Jelliffe unveiled how capitalist enterprise could generate hunger.[116] McLaren provided a concrete example in his writings on the fiasco, where he observed that bottle feeding in unhygienic conditions actually caused marasmus.[117] These critiques were taken up by organizations such as War on Want in the United Kingdom, whose "Baby Killers" report inspired the Nestlé boycott and became a catalyst for decades of organizing against a capitalist food system that produced wealth for the few and hunger for the many.[118]

Eventually these critical voices gained traction, international agencies started to recognize the limits of protein supplementation, and understandings of the relative role of protein in global malnutrition began to shift. Revised clinical assessments of kwashiorkor patients found them to be deficient in calories as well as protein (a status that became known as "protein-energy

FIGURE 2.2. A World Food Conference under the UN General Assembly held in Rome from November 5 to 16, 1974, addressed the issue of global famine and hunger as the tide began to turn against protein primacy. UN Photo/F. Lovino.

malnutrition") and a series of papers was published suggesting that smaller amounts of protein were necessary to prevent and treat kwashiorkor in infants than previously thought. In response, the UN slashed its recommended protein allowances, "unwittingly" closing the protein gap in the process.[119] It wasn't until July 1974, however, when the *Lancet* published "The Great Protein Fiasco," that the need for radical change was widely and publicly acknowledged.[120] The timing of McLaren's argument was crucial: Despite intense efforts to address the so-called protein crisis over the preceding decades, more children were suffering from malnutrition "than ever before."[121] This situation had unfolded even as considerable resources had been wasted on research projects, conferences, and food experimentation and development, McLaren contended, sarcastically suggesting that "instead of cost/benefit analysis, someone should calculate a cost/detriment analysis."[122]

Amid turf battles, infighting, and increasing evidence that a change in direction was required, not to mention a famine in Bangladesh that had killed an estimated 1.5 million people by 1974, the WHO decided that PAG had out-lived its usefulness, and the group was disbanded. The following year, the UN's Advisory Committee on Coordination met to determine why, "'despite overwhelming moral imperatives,' the UN and partner governments had not produced a strategy for the elimination of hunger and malnutrition."[123] One answer is that questions raised about protein and kwashiorkor had "simply become a justification for more research," in Scott-Smith's framing of the problem, a point McLaren took further in the *Lancet*, where he claimed that the nutrition-industrial establishment had blocked any attempt to widen the focus beyond protein. Scott-Smith goes on to note that even John Brock, coauthor with Autret of the influential 1952 report, admitted in his own 1974 *Lancet* article that human needs were not driving research: "We could have described many undernourished and marasmic African children," Brock said, "as they were all around us. But they did not represent what we had come to study."[124] One year later, leading advocate of the protein problem Dr. John Waterlow coauthored an article in *Nature* titled "The Protein Gap," which acknowledged that no gap existed and that sufficient energy intake was all that was required for children in developing countries to be adequately nourished.[125] The real problem, experts now claimed, was food supply. As a result, the very idea of protein malnutrition was "suddenly discarded," and other orthodoxies (namely a focus on micronutrient malnutrition) emerged in its wake.

To put it bluntly, the efforts of the UN and its member governments to address malnutrition through a focus on protein supplementation were a dismal failure, laying bare the mechanisms by which people's nutritional needs are shaped by geopolitical power. Taking such mechanisms as our lens, the story might be summarized this way: The rapid intensification of Western agricultural production in the first half of the twentieth century led to market yields that exceeded what was necessary to build the strong colonial bodies that would capture, expand, and defend colonial territories. The growth of industrial agriculture was predicated on the decimation of Indigenous food systems and ecologies both at home in the United States (through cattle colonialism) and abroad. The food insecurity that resulted in turn built demand for nutritional supplements derived from non-Indigenous, energy-intensive, industrialized food sources that usually cost more to produce than the local crops they were designed to replace.[126] With a long-standing reputation as a "charismatic nutrient," protein deficiency was readily identified by researchers as the primary driver of malnutrition, though clear evidence for this supposition never emerged. The singular focus on protein was compounded by a rapidly expanding nutrition science culture in which researchers chase grant money, covet seats at the global policy table, and seek to "'make a case'" for the importance of "their nutrient."[127] It was also a moment when optimism about the promise of techno-capitalist solutions for resolving world problems was running high, and protein in its molecular, high-modernist form was a readily adhesive partner. This is not to say that there weren't children suffering from inadequate protein intake but, rather, that protein deficiency was not an independent cause of malnutrition across the Global South, as protein crisis proponents claimed. Protein deficiency was construed as such largely because of Western culinary and corporeal norms rooted in racist assumptions and arrogance, not reliable evidence.

Conclusion

In 2000, twenty-six years after his historic intervention, McLaren revisited the fiasco. Claiming that "PEM [protein-energy malnutrition] is now recognized to be a complex, deep-seated, and massive public-health problem that will respond only to general development," he declared with some satisfaction that the fact that "the protein-rich food mixtures and all the other trappings of the fiasco have long disappeared is evidence that that particular fallacy is long dead and gone."[128] While there is general consensus that the period from the mid-1970s to the present has been dominated by a micronutrient paradigm,[129] McLaren's assessment seems optimistic and sweeping from today's perspective. As Richard Semba explains, the intense focus on the "protein gap" may have diminished after the mid-1970s, but expert committees of the FAO and the WHO have continued to address protein requirements and quality, issuing reports in 1973, 1985, 2007, and 2013.[130] Moreover, the science of protein in relation to malnutrition is far from settled. "Recent studies challenge the widespread assumption that young children in developing countries receive sufficient dietary protein," Semba notes, going on to lament that almost fifty years after the Great Protein Fiasco concluded, "the protein and amino acid needs for children in the first 3 years of life, or the first 1000 days when children are most vulnerable to stunting, remain poorly understood."[131] Semba is part of a revived push to reconsider the role of protein and other facets of food quality in "stunting," specifically the efficacy of milk and meat because of their "strong association with linear growth in children in developing countries." Semba also points to the "renewed emphasis" on technological solutions to protein malnutrition, a shift that is part of a broader return to food supplementation in development projects.[132]

At the start of the 1990s, the harmful impact of structural adjustment policies and the reduction of state-financed nutrition programs on malnutrition rates was increasingly evident. Nott notes that inpatient mortality for malnutrition stood between 20 and 30 percent at this time—the same rate as in the 1950s. This sorry situation gave "nutritionists renewed purpose and space to operate," Nott claims, and a rash of new treatments emerged.[133] For a variety of reasons, including that they don't require refrigeration and can be used without admission to a hospital, these new treatments have helped slash mortality rates. Plumpy'nut, a "protein packed" oil-based paste developed in France, is the most prominent and widely acclaimed of these foods. Tom Scott-Smith is among the critics of this "miracle" treatment and "magic commodity," noting that it does nothing to address the structural injustices at the root of poverty and malnutrition and that it facilitates the increased involvement of the private sector, facilitated by the UN, in the management of malnutrition through engineered foods.[134]

Technofixes aside, nutrition scientists and international agencies remain generally cautious about overemphasizing the role of protein in malnutrition. Yet true to form, protein has found other agendas to which it can adhere. In the years since the idea of a protein gap lost steam, debates about appropriate protein intake among the satiated classes in the Global North have emerged in its wake. As concerns about chronic disease and an individualist emphasis on managing one's lifestyle and wellness have risen to the fore, so too has the marketing of nutritional fixes. In the 1970s, protein began to appear in health food stores, often in the form of individual amino acids ascribed with properties that might lead to fat loss, renewed energy, or resistance to aging.[135] Frances Moore Lappe's best-selling book, *Diet for a Small Planet*, also helped drive sales.[136] Here Lappe made an argument for environmental vegetarianism, but because she feared that readers would worry about insufficient protein if they were to follow such a diet, she made an argument for combining proteins, by which eaters must attend carefully to the particular blend of plant foods consumed at individual meals to ensure optimum protein intake. While Lappe later changed her mind about the need to adopt a scientific approach to protein combination, acknowledging instead that eating a variety of vegetable foods throughout the day would provide adequate amino acids, her book had the effect of driving a focus on protein and protein supplementation among the many followers who were inspired by her work.

By the 1990s, according to Carpenter, the major share of amino acid supplement consumers were bodybuilders, many of whom were inspired by advertisements in muscle magazines espousing the anabolic or tissue-forming properties of their wares.[137] As we discuss in chapter 4, new disease categories linked to the idea of protein deficiency among relatively affluent Westerners have emerged since the 1990s; the most prominent of these is sarcopenia, or loss of muscle mass with aging. And as we explore in chapter 5, the same period has seen protein supplement use among fitness enthusiasts and athletes spread far beyond bodybuilding to a much broader range of health, food, and fitness cultures, albeit with an enduring homology between protein, muscles, and masculinity leading the way. These shifts have helped generate fresh markets for high-protein foods and unlock profitable avenues of knowledge production for scientists across a range of fields beyond global nutrition.

As these chapters will attest, despite the Great Protein Fiasco, despite the "protein gap" being labeled nutritional science's biggest error, and despite heightened awareness of the environmental implications of many high-protein foodstuffs, today we are once again living through an age of protein boosterism. Protein-centric thinking hasn't disappeared in the wake of these errors and controversies, but is instead thriving in "protein cultures" such as the low-carb diet movement and sport science research, where protein's role in bolstering athletic bodies remains a leading area of interest. Thus, contra to extant histories, we suggest that protein flourishes *because* of fiascos and gaps, not despite them. This is a story not of a mistake but of a shape-shifting continuity by which protein's allure adapts to new contexts, helping reproduce globally stratified and physically embodied dichotomies between healthy and unhealthy bodies—in concert with responsibilizing discourses of development and lifestyle.

3

FROM GUTTER TO GOLD

A Political Ecology of the Protein Powder Industry

These . . . components of whey, notably the proteins and peptides, and their properties . . . have helped transform whey from a waste material that has often been shunned, to a valuable dairy stream containing a multitude of components available for exploitation in the agri-food, biotechnology, medical, and related markets.—Geoffrey Smithers, "Whey and Whey Proteins—From 'Gutter-to-Gold'" (2008)

This chapter travels across the Atlantic to midcentury North America, where the rapid industrialization of dairy agriculture and its ecological fall-out played leading roles in the commodification of whey—by far the most common source of protein supplements. Beginning in this place at this time means skipping over the vast history and countless encounters and innovations hitherto shared by humans and whey. Humans have been making use of this coagulated liquid as pig feed, fertilizer, and food for about eight thousand years, or at least since it was noticed to separate naturally from milk and cheese in the Fertile Crescent.[1] Whey in this curdled, semisolid state has been a constant in what is known of human civilization, its dense capacities having been put to work in artisanal recipes and medicinal potions long before its contemporary manifestation as a health and fitness supplement. But whey's remarkable transmogrification into the powdered form with which it is now synonymous occurred relatively recently, in the postwar United States.

A storyline is already appended to these changes, one that loosely follows from food industry consultant Geoffrey Smithers's summation in our epigraph. It goes something like this: *Faced with enormous quantities of toxic whey effluent generated by intensive dairy production during a period of accelerated agricultural development, capital investment and technoscientific ingenuity responded as one to address whey as an environmental problem, resulting not only in a solution to those problems but a multibillion-dollar protein powder market.* In broad strokes, this is whey's official biography, and as one might expect given the cast of agribusiness enthusiasts who tell it, it is a triumphant tale of whey ascending from problem to product out of the tumult of dairy industrialization. The environmental problem of whey is left behind as quickly as it is invoked, serving only as a plot point that emphasizes this remarkable achievement. Every whey-infused smoothie, shake, and bar that makes up the contemporary protein powder market serves as a testament to this success.

Our version of whey's twentieth-century biography diverges from the established rendition. While we work within its general parameters, we do so with a sensitivity to protein's elusiveness, dynamism, and multiplicity that casts this story in a different light. Whey has indeed charted a path from noxious effluent to nutritional commodity over the past half century, but its trajectory has been shaped by a messy ancestry of techno-innovations, industrial agricultural excess, capitalist imperatives, dairy cow exploitation, environmental campaigning, and a broader biopolitical shift toward the optimization of bodily health. In place of a simple ascendance from agrarian excess to pollutant to commodity, we find a less linear journey, one in which whey's potency is the source of both its vital power and its deadly effects. As we will see, even where that potency is harnessed toward productive physical and economic ends, whey is never quite purified of its past, even once it is ingested. Instead, those qualities are deferred and displaced through the bodies of those who consume it, who must undertake certain kinds of athletic and metabolic work in order to process this substance and realize its promises. This metabolic labor, we suggest, is seldom figured in theorizations of social metabolism, just as socio-ecological questions and concerns almost never find their way into dominant understandings of metabolism as an individuated, intrabodily process.

In this chapter we attend critically to whey's production, ingestion, metabolization, and excretion, tracking technoscientific attempts to manage the unruly and environmentally destructive proteins by which it is constituted. In doing so we confound whey's official biography by attending to its *capacities of refusal* in biological and ecological systems that seek to purify its toxicity and profit from its potential. This means charting how protein travels and mutates through and with bodies and machines in a dance of dispersion and assemblage that creates surpluses of value and waste, strengthening some bodies and compromising others along the way. Through discussions of its nebulous character as both noxious waste and hyper-refined source of value, its emergence through crises of milk and cheese production, and its lessons for understanding the socio-ecological embeddedness of protein metabolism, we argue that whey's evasion of total capture in this complex choreography is one of its defining characteristics. It refuses to acquiesce entirely to the attempts of agribusiness and biomedical professionals, as well as those of consumers, to harness its vitality, and refuses to be purified entirely of the nitrogen density that make it so potent in the first place. As such, whey is exemplary of the complex constellation of economic, ecological, technoscientific, governmental, and affective-corporeal dimensions involved in protein's broader cultural ascendance. To see how, let us begin at the end of this convoluted set of processes, with whey in its commodified, hyper-refined form.

Whey, Waste, and (Bio)Value

The image of whey protein most familiar to consumers is of a magical powder endowed with abundant benefits. Fitness instructors, exercise scientists, dietary influencers, protein marketers, and dairy industry lobbyists strike a chorus of praise about the wonders of protein powder consumption, which are proclaimed to range from enhanced musculature and lower blood pressure to glossy hair and sharpened acuity. Such is the power imbued in this elixir that even seemingly unhealthy products such as beer and ice cream now come infused with whey protein in expensive artisanal concoctions. We venture that if you have not had the virtues of protein powder extolled to you by a health and fitness professional, you have almost certainly been targeted by food marketing and labeling in which the importance of protein is conveyed in grams, servings, and desirable images of vitality.

While all sorts of expansive claims are routinely made about protein writ large, protein powders are an increasingly common means of consuming

FIGURE 3.1. Protein supplements are commonly available in powdered form and are typically packaged in large plastic tubs, reflecting their widespread use in fitness and nutrition contexts. M. Verch. CC BY 2.0.

protein—even for those who may not realize that their snack or beverage is infused with this substance—and whey powder is the most popular variant in this saturated market. Promotional claims may take the form of recommendations from a physician or coach, click-friendly lists on health and exercise websites, or text on the plastic tubs of powder and whey-infused products that line supermarket shelves and fill online shopping carts. Some of the hype around whey's benefits can of course be attributed to the logic of advertising that pervades consumer capitalist societies, which makes particular demands of different populations in relation to food.[2] Yet the discourse promoting whey protein reaches across the commercial and the scientific, the popular and the prescriptive, and is at least superficially consistent in its content and reverential tone, blurring the ever-fuzzy borders between health, fitness, and consumption. This means that, if it is debated at all, whey's value to the bodies and minds of its consumers is considered almost entirely in degrees of magnitude, in terms of just how much it can enhance those who ingest it. Where questions do arise, they tend to concern which combinations of bodily labor, exercise habits, and concomitantly ingested foodstuffs might best draw out whey's fortifying properties. Whey's status as a health elixir, a nearly ubiquitous reputation within health and exercise cultures and (sport) scientific epistemologies, is not troubled by this discourse.

To speak of whey protein as waste, then, is to diverge starkly from prevailing associations with health, vitality, and value of different kinds. The material form this commodity takes is a starting point for understanding why whey's status as a waste product seems so counterintuitive. Whey protein immediately defies associations with waste through its appearance as a filtrated, concentrated, desiccated white powder: a granular, molecularized form ostensibly free of all that lacks use or purpose. It remains for us a curiosity that this powdery substance has become so synonymous with liveliness and vigor despite its appearance as dead, inert matter, perhaps as far from vitality and health as one could render or indeed imagine any foodstuff. Take, for example, two forms of whey powder that are presented as contrasting products in the marketplace, whey "isolate" and whey "concentrate." Whey isolate goes through more intensive processing than whey concentrate, which is understood to reduce the fat and carbohydrate content of the former, and typically to make it more expensive. But in each case, whether isolate or concentrate, the inference is that whey is processed to eliminate excess matter and that the more refined the processing, the purer the product—unless we count the plastic tubs.

Scientific accounts of whey's effects within the body bolster this impression of wastelessness and of hyper-refinement. Whey scores highly in various protein assessment scales, including its biological value (how efficiently the body absorbs nitrogen from a particular protein source) and the FAO's current metric, the Digestible Indispensable Amino Acid Score (how efficiently the small intestine absorbs amino acids). The high ranking of whey protein in these metrics translates into an especially valuable quality for consumers: the speed of metabolization. Whey is listed among the quickest forms of protein to metabolize, indeed the quickest in several of these ranking systems. The benefits of speedy replenishment to athletes and bodybuilders are fairly obvious: The sooner one metabolizes protein, as exercise science tells it, the sooner one is repairing and regenerating muscle tissue damaged through intensive exercise. But there is a wider, more diffuse appeal of speed for consumers who may not be athletes or even fitness aficionados but who are increasingly used to time being measured in terms of efficiency and productivity, for whom speed itself signals value. By this measure, whey sits in contradistinction to notions of waste in part because it seems to waste no time in realizing its potential. Unlocking its resplendent promises is accordingly simple and swift: Just purchase foodstuffs already infused with whey or add it yourself to shakes or smoothies, mix consumption with athletic labor or "exercise"—or any calorie-burning activity, depending on whose advice you take—and the benefits will soon be yours to reap.

When metabolic speed is added to whey protein powder's visual and sensory qualities of hyper-refinement, its status as the foremost deliverer of protein's über-nutrient potential becomes easy to understand. There are some caveats to be found if one dives into scientific debates about protein metabolization, where questions about what else was consumed alongside protein, the sleeping patterns of the individual in question, and matters of urea excretion inform attempts to measure the "true biological value" of protein. Alluring as this quest for truth may sound, our own epistemological leanings teach us that digging deeper into essences to excavate true value is less revealing than expanding one's analysis out to apprehend the social conditions in which some things are constituted as valuable and others as waste.

Thinking about the social treatment of waste in consumer capitalist societies aids an understanding of whey protein as a multiplicitous substance that variously manifests as valuable or wasteful, productive or polluting, depending on the context of its materialization. Waste calls to mind something worthless and thus discarded, whether fast fashion, digested food, or some other object suspended between usefulness and nothingness. In industrial societies, waste is hidden away in containers, pipes, treatment plants, and landfill sites wherever possible, and spills into sensory proximity only when social processes of erasure are disrupted or exposed, or when efforts to maintain its invisibility do the opposite. This latter point is captured by John Scanlan in *On Garbage*, where he gives the example of how the "great cleanup" of the nineteenth century, which forced food retailers to protect food from contamination and spoiling by utilizing new forms of packaging and storage, "paradoxically creates more material garbage, which in turn constitutes part of the greater problem of environmental degradation that we are told threatens life on a far larger scale."[3]

Mary Douglas's anthropological insights about waste are also called to mind here. Douglas famously argued (via William James) that the constitution of dirt is culturally variable and contingent on social norms and cues, such that dirt is less a verifiable essence, wherever it appears, and more "matter out of place."[4] Similarly, theorist of science and technology Bruno Latour called this convoluted process of sorting matter "purification" and identified it as a quintessential characteristic of modernity. In adhering to the fantasy that the world can be neatly carved up and contained in dual categories of pristine nature and man-made society, or in this case clean, "pure," decontaminated food and the dirty processes of its production, modern societies increasingly create "hybrid" entities (e.g., plastic packaged food) that blur these boundaries further and proliferate their contradictions.[5] And in the twenty-first century, when these hybrids are reproduced at an unprecedented rate and scale, maintaining this semblance of invisibility is increasingly important and ever more difficult. Whey thought through this schema is "purified" not because of its inherent purity or any aesthetic or biochemical qualities but, as we will soon see, because of the extraordinary and never-quite-complete attempts to strip it of its inherent toxicity and from the multiple lifeworlds implicated in its creation.

Thinking about waste as an inevitable outcome of efforts to purify commodified food of its production processes takes us toward an understanding of whey as more than a biochemical property or nutrient. In the process, whey reminds us that waste is not in fact a fixed category, does not necessarily lack worth, and does not indicate the end of an object's life cycle. Viewed over a longer historical period, and with a socio-ecological sensibility toward understanding its materialization, we can see how whey confounds distinctions between what constitutes waste and what constitutes value. This is vividly illustrated by whey's materialization as a hazardous outcome of dairy overproduction and its constitution as a waste product, which together formed the impetus for its twentieth-century transformation.

Got (Too Much) Milk? Resolving Crises of Dairy Overproduction

In his ethnography of hog factory farming in the United States, Alex Blanchette notes that since the 1990s "agricultural colleges have tried to invent schemes that would turn [hog] waste into the 'the new black gold (i.e. petroleum).'"[6] Hog fecal matter pools in the vast "manure lagoons" that result from industrial-scale raising, feeding, and killing of hogs, creating more semitreated waste than the sum of 50 million people. Evidence is growing that these craters of animal waste, hundreds of which mark the landscape of the Great Plains and Midwest of the United States, become airborne in high winds and can even "affect regional microbial ecologies through horizontal gene transfer," spreading not just the obvious odors and particles but also, potentially, antibiotic resistance.[7] That this is framed as an "environmental problem" begetting research and investment rather than a consequence of tearing asunder the balance between crops and cattle as "arbiters of each other's growth" exemplifies how industrialized agriculture positions itself as a provider of solutions to the problems it has wrought.[8] And in this case, as so often, the solution centers on rendering something of economic value from the hazardous wastelands of hog farming.

For inspiration, those agricultural colleges charged with turning hog waste into the "new black gold" might look to the response of dairy farms to their own waste production in midcentury America. They would encounter there a similar story of the ecological damage resulting from industrial-scale dairy production and, if adhering to industry-sponsored research and reports about whey, would learn of its ostensibly triumphant ending in the form of a market for whey protein currently estimated at around US$10.26 billion, narrated by industry insiders and corporate researchers such as Geoffrey Smithers as an ascendance from "gutter-to-gold."[9]

Much like the advent of the advertising and financial credit systems that took off in the early twentieth century, many food production innovations at this time were intended to avert capitalist crises of overproduction. The yields of industrialization required demand for goods to match an unprecedented scale of supply, and nowhere was this more evident than in the growth of milk in the American diet. Demand for canned milk, in particular, had skyrocketed during World War I, when dairy farmers had ramped up production to provide soldiers with sustenance overseas. After the war, the demand for milk no longer reached those heights, and dairy farmers were left with enormous surpluses and high production costs. Responding to this economic downturn and to the threats of dairy farmers to organize unions in the 1930s, federal programs were set up to intervene. In 1946, for example, the National School Lunch Act was passed by the US Congress. This Federal Law proclaimed fears about poor childhood nutrition, and part of its remit was to mandate that every school lunch include "between one and a half to two pints of milk." In this act, biopolitical anxieties about the vitality of future populations merged with economic interests and concerns in the shape of a nationwide intervention in children's nutritional intake. Despite this and other ensuing government interventions, including a boom in instant milk powder production in the 1950s discussed in the previous chapter, human consumption of cow's milk reached its peak in the United States during World War II and has been in decline ever since.[10]

This downward trend in milk consumption led to whey's incarnation as powder, albeit via a convoluted path. Health concerns about milk have their own history owing to its high susceptibility to bacterial growth. The introduction of pasteurization in the early 1900s played a decisive role not only in ridding milk of certain pathogens but also in extending its shelf life, and in turn expanding the operable distances between dairy production and consumption. The healthiness of milk consumption nevertheless remained contested through twentieth-century fluctuations in demand. Heightened public awareness that milk is nutritious for calves but hard to digest and nutritionally unnecessary for humans, not just among those who are diagnosed as lactose intolerant, has spread widely in the United States and beyond. Specific concerns surrounded the health effects of feeding hormones to animals that are then ingested by the humans who eat them, something that is endemic and well documented within the factory farming system. These concerns are buttressed by the range of other, "alternative" protein sources now available to those put off by cow's milk. And compounding each of these factors is the animal rights advocacy that has challenged the mass incarceration, torture, and slaughter of cows used in industrial milk production and consumption. Exemplifying this shifting nutritional terrain, First Lady Michelle Obama's 2012 intervention in school lunches—an intervention that placed restrictions on particular food groups, including milk fat, that were categorized as staples in the National School Lunch Act of 1946—marked a stark departure from milk's midcentury heyday.

Through this crisis of milk manufacture, we see how political will drives capitalism's propensity to create markets that ostensibly mitigate its excesses—something that continues into the twenty-first century, as the US government now subsidizes a dairy industry that operates at a substantial loss.[11] In 1995, Dairy Management Inc. was formed by the National Dairy Promotion and Research Board and the United Dairy Industry Association, a conglomeration in the making for many decades. Dairy Management Inc.'s funding comes from the US Department of Agriculture, and its mission is unequivocal: to create demand for dairy products. As journalist Michael Moss documented in a 2010 exposé in the *New York Times*, one of Dairy Management Inc.'s great successes in its previous iterations had been finding a home for milk surpluses in the form of cheese.[12] The campaign to encourage people to eat more cheese has been astoundingly successful: In 2016, Americans ate an average of 17.46 kilograms (38.5 pounds) of cheese per capita each year and Canadians 13.46 kilograms (29.7 pounds), almost triple the amount consumed in 1970.[13] This shift has in turn been profitable, given the higher price commanded by cheese compared with milk, helping dairy farmers stay afloat in the face of fluctuating markets.[14] Yet again, though, the solution begets a problem: Because only 10 percent of milk can be turned into curds, the shift from milk to cheese created an enormous surplus of whey effluent.

The numbers make for stark reading. Nine liters of whey remain for every kilogram of cheese produced, and a large cheese processing plant can produce over 1 million liters of this protein- and acid-rich substance daily.[15] In 2006, the US dairy industry generated 90.5 billion pounds of whey effluent.[16] By 2017, as the *Washington Times* declared, "cheese stockpiling has just hit an all-time high": 1.39 billion pounds of cheese were being produced at a rate that US domestic consumers and export channels together could not match.[17]

It's worth pausing here to remember that whey has always accumulated as excess, and the time-honored responses of farmers has been to reuse whey "waste" materials such that they were not categorized as waste at all. For a comparative example that anticipates the fate of "excess" liquid whey, take Hannah Landecker's analysis of beet pulp, which also centers on the early twentieth-century United States. Beet pulp is the excess amassed from beet sugar production once the sugar itself has been removed. Beet pulp became a waste product and an odiferous pollutant as sugar production rates grew with industrialization. What to do with more than half a million tons of wet beet pulp that has served its purpose, with the sugar long since departed to American stores, homes, and stomachs? The solution to this problem came in the form of drying techniques developed in Europe, through which pulp was desiccated and therefore made more easily transportable and efficient as animal feed. Fertilizer and feed had been the purposes served by such waste products for centuries, but the hauling costs and sheer amounts of beet sugar produced in industrialized systems had rendered this harmonious relation outmoded. In its dried form, however, the "feed efficiency" of sugar processing operations was given an enormous boost, increasing the production of goods and profit rates.[18]

As with beet pulp, whey did not pose a problem to dairy farmers before the industrial transformation of dairy farming, since it was mainly used as hog or chicken feed if not sold to or traded with neighboring farmers, and some farmers would raise hogs expressly for this regenerative purpose. But the transfer of manufacture off the farm into factory systems, and the turn to cheese-making, rendered this maneuver outmoded. Reuse of liquid whey was not only logistically problematic but also expensive, requiring diesel-fueled tankers to haul heat-sensitive whey back to factory farms to the animals who would consume it following treatment. The question thus became: What to do with the abundant whey effluent that is a necessary by-product of cheese manufacturing, but not itself the desired product?

FIGURE 3.2. In June 1950, a transport truck in Appleton, Wisconsin, picks up whey from cheese factories. This is an early example of efforts to reclaim and reuse whey that had been previously dumped as waste, thus causing stream pollution. F. H. King, Wisconsin's Historic Natural Resources Photos, State of Wisconsin Collection, University of Wisconsin–Madison. CC BY-ND 4.0.

As Scott Lougheed has shown, the answer initially took the form of pollution. In the 1960s and 1970s, US cheese manufacturers set about whey dumping, spreading whey effluent on agricultural land, pumping it into municipal sewers, or simply draining it in public water systems.[19] Due to its high nitrogen content, whey is 175 times more potent than untreated human sewage.[20] Whey dumping thereby became a major source of water pollution in cheese-making regions such as Vermont, Wisconsin, and Ontario in the mid-twentieth century.[21] Unsurprisingly, this proved ecologically devastating and untenable. Whey's high nitrogen content spurred plant growth in waterways newly awash with dairy effluent, and multitudes of aquatic life died from lack of oxygen on account of its stifling presence. Fish kills and contaminated soils were the upshot of whey's proliferation in these water systems.[22]

Widespread whey dumping was quickly followed by various efforts to highlight and oppose its catastrophic effects. The smell alone drew the ire of communities whose local rivers and streams had become dumping grounds in concert with the boom in cheese manufacture. Throughout the 1970s, a wave of environmental activism and calls for legislation across the United States and Canada prompted governments, cheese-makers, and food industry researchers, to respond urgently to the problem of whey waste.[23] Something had to be done, and if making less cheese was not a palatable alternative—at least not among the assembled crowd of change-makers—then that something meant a repurposing of whey effluent, the creation of a new market for this centuries-old foodstuff that had seldom been less than a dexterous ally to dairy farmers of yore.

It is at this juncture that the celebratory story of whey protein begins its ascendance, that the official biography reaches its moment of redemption via repurposing. In response to environmental activism, nascent legislation, and the persistent challenge of resolving the crises of milk and cheese overproduction, technoscientific innovations in the filtration, concentration, and desiccation of liquid whey made significant gains, eventually resulting in a powder that was easier to store, more agile to move, and, crucially, more digestible and appetizing for humans. In place of a substance of renowned "toxicity," a liquid pollutant of marginal worth that had spilled into the lives of people and animals with harmful consequences, this newly forged granular substance represented uncharted futures for whey and an extraordinary break with thousands of years of its mostly stable, unremarkable role as food for humans and livestock. Protein powder as we know it emerged from this constellation of economic imperatives and ecological fallouts. These, too, were the necessary conditions for any notion of whey's biological value to be conceivable, and eventually the basis of its central role in nutritional and exercise science and "nutricentric" body cultures.[24] Hence the narrative trajectory of whey's arc from gutter to gold, denoting its journey from harmful toxin disposed in rivers, streams, and sewers into "a valuable dairy stream containing a multitude of components available for exploitation in the agri-food, biotechnology, medical and related markets."[25]

Perhaps the most curious feature of this story is not whey's upward trajectory from gutter to gold—after all, the analogy came from a researcher who declared funding from the dairy industry, and the desirable "end game" is easily lauded after the fact. Nor is it the celebration of a market for whey as a victory in itself, given that markets have become what Achille Mbembe has called "the most undisputed forces of our times" and that the protein powder market is far from unique in this respect.[26] Rather, it is that whey's official biography troubles any notion that the growth of a protein powder market is first and foremost a cultural phenomenon, indicative of what Nikolas Rose might call an "ethopolitical" desire for individual health and body optimization.[27] This is implied by those who have observed that demand for protein powder has burgeoned out of bodybuilding subcultures in the 1980s and 1990s into a multitude of health and fitness regimens, a classic case of post hoc logic that is internally valid yet misses out on the significance of socio-ecological materializations in the making of protein powder as a substance that is conducive to such wide markets.

More curious still is that the storying of whey's commodification would be recounted in much the same way by staunch supporters and vehement critics of dairy industrialization and the environmental impacts of capitalism. The difference is in the emphasis placed on whether this is indeed a triumph or yet more evidence of capitalism's propensity to produce and only temporarily resolve or "fix" crises in a dialectical process it is doomed to repeat. The upshot is that any critique of these crises of dairy overproduction that leaves intact the integrity of the purification of whey and its ascendance from gutter to gold meets its limits as an aversion in principle to commodification or capitalism. The fabled process itself, the journey from noxious effluent to health and fitness commodity, goes untroubled—unless we continue to follow our subject beyond the reified boundaries of its commodification and into the biochemical processes of consumption, digestion, and metabolic processing.

Metabolism as Waste Disposal Made Flesh

Metabolism is an interesting word. Much like *protein*, it most often refers to an intrabodily phenomenon whereby foodstuff becomes energy. In dietary cultures, metabolism is shorthand for how quickly and effectively one can process the things one has eaten. But it's so much more than that.

Take a moment to imagine yourself sipping on a whey smoothie, eating a protein-infused snack, or unwittingly consuming a product that has been fortified with protein powder. That moment of ingestion is a transitory coalescence of socio-ecological relations: a coming-together of technosphere, biosphere, and ecosphere that marks a kind of incarnation of the story narrated above. This is always the case, as the dictum "We are what we eat" reminds us, and the details matter: where, when, and with what accompanying foodstuffs or bodily labor makes a difference in whether and how protein releases its qualities. Metabolism tends to denote what happens after this moment of ingestion in referring to the chemical reactions that are catalyzed by consumption.

Talk of protein metabolism in exercise science and fitness circles follows this logic and, accordingly, is mostly about regime and efficiency. Through what programs of eating and movement can the body optimally metabolize its protein intake? Countless research articles, blogs, threads, magazines, and other mergings of exercise science and popular culture will offer answers to this question, often in the form of prescriptions (for kinds of protein), regimes (of physical activity), and a litany of consumer choices gathered under the sign of "lifestyle." This is metabolism as a biochemical process to be mastered through scientific knowledge and, crucially, through the will to subject oneself to dietary and fitness-related prescriptions for healthy living.

There is no doubt that metabolism is a dynamic process of breaking down and regenerating substances that is fundamental to sustaining life. Look more closely, though, and uncertainties and contingencies appear. For example, as we discussed in chapter 1, what exactly happens in the folding of proteins from polypeptide chains of amino acids into the bewilderingly complex structures called proteins is not comprehensively understood, despite its centrality to metabolic processes. The breaking down of bonds that hold amino acid chains together (catabolism) and redeployment of those chains for new purposes (anabolism) are comparatively well-established processes, but even these are matters of ongoing debate and contestation. The speed of protein metabolism, for example, is also influenced by myriad factors in the chewing, swallowing, and digestion of proteinous food. As it turns out, a surprising amount of the what, why, and how much of protein metabolism is fraught or contested beneath the surface of a consensus of convenience between knowledge about metabolic processes and those who seek to benefit from this knowledge in myriad ways. Without figuring this complexity, a machinic image of the metabolizing body prevails, a conception of bodies as kinds of biochemical processing sites through which "proteins" are synthesized or excreted. This is metabolism's prevailing biochemical and discursive form, whereby fast and slow processing facilitates the making of more and less efficacious, and more or less desirable, bodies. And it is understood as happening almost exclusively *within* the biological body.

Other understandings of metabolism are available and, we contend, are necessary for grasping the socio-ecological processes inherent in the consumption of this particular variant of protein, whey powder, at this particular moment in its history. In social theories with environmentalist bents, metabolism has become a leading conceptual tool for analyzing the relationship between capitalist economies and ecologies of various scales. Karl Marx's ideas are often the starting point for these analyses. Marx wrote of an "irreparable rift" in the social metabolism (*Stoffwechsel*) brought about by capitalist agricultural production, in particular the depletion of soil that comes with moving agriculture away from peasant farmers into capitalist family holdings and creating spatial distance between sites of production and consumption.[28] This is a "metabolic rift" insofar as it disrupts the equilibrium of humans and natural systems, bringing about ecological crises that subsequent modes of agriculture have been charged with resolving. John Bellamy Foster revived this idea in *Marx's Ecology: Materialism and Nature* (2000) as part of a boom in environmental thinking in the 1990s that wedded Green and Red thought in a timely synthesis. Foster mounted a dual critique of a pervading spiritualism in environmentalist movements, on the one hand, and accusations that Marx lacked an ecological sensibility in his analysis of capitalism, on the other. The concept of the metabolic rift was crucial to advancing both strands of his argument: In its focus on earthly systems and substances such as soil, it demonstrated the material basis of the disruption to the natural world wrought by the onset of capitalism and in doing so heralded Marxism as a persistent resource for environmentalist theories and movements.

Openly indebted to the Green Marxism championed by Foster while positing its limitations, Jason Moore's world-ecological approach argues for a metabolic understanding of capitalism and nature that does not reify these two as separate realms, the former doing great violence to the latter.[29] Moore is concerned that an ontological distinction between Nature and Society is reified through the idea of the metabolic rift, as if capitalism was an external system imposed onto the natural world. More than an intellectual impasse, Moore follows strands of feminist, decolonial, and Indigenous thought in seeing this dualistic conception of nature and society as inseparable from the violence inherent in the making of the modern world. After all, it is capitalism that abstracts the nature of the body into wage labor and exchange value, colonialism that instantiated this value system through enslavement and located originary nature within particular people and places, and orthodoxies in modern science that aided in authenticating these abstractions such that they became naturalized as common sense. The metabolic rift, then, conceptually reinscribes and reproduces these epistemic dualities. Instead, Moore offers the *metabolic shift*: metabolism as a flow of capital, power, and material nature in which capitalism and ecologies are coproduced in shifting configurations of human and extrahuman nature. Instead of a rift between nature and society enacted by the onset of capitalism, think of flows through worlds that are irreparably both natural and social, immanent to each other, where what we constitute as nature responds to its reorganization. Capitalism and other world-historical processes emerge not as impositions onto nature that cause rifts in hitherto harmonious ecospheric and biospheric systems, but as attempts to organize natural systems, which become increasingly unruly in the process. As such, these attempts are seldom simple and efficacious. In Moore's words, "While the manifold projects of capitalism, empire, and science are busy making Nature with a capital 'N'—external, controllable, reducible—the web of life is busy shuffling about the biological and geological conditions of capitalism's process."[30] Put differently, forces aimed at purifying and profiting from nature meet *with* worlds that are themselves composed of active and reactive forces, all of which makes up what Moore calls the web of life. Substituting *protein* for *nature* in this passage allows us to grasp its persistent "unruliness" in the face of two centuries of scientific, colonial, and capitalist efforts at mastery and governance.

There are always simplifications and erasures when one constructs the lineage of a concept in the way we have here. It bears emphasis that notions of metabolism travel under different signs within and beyond the shadows of Marxist thought, not least within the contemporary life sciences, and these too figure into thinking about protein. For just one example, in her theorizing of the "new metabolism," Hannah Landecker finds in the field of nutritional epigenetics an understanding of food as "a form of environmental exposure" wherein what we eat is understood as an interface between our genes and our environment.[31] This relational understanding of biology and society is not present in Foster's tumultuous dialectic between capitalism and nature—the rift—or Moore's immanent dialectics of flows with and between socio-natural processes. But they each share this same relational premise that takes metabolism *out of the body* to understand it as a dynamic worldly process.

With Foster, Moore, and Landecker as especially illuminating guides, we can see that protein metabolism is not simply an intrabodily process, or at least not only that. Protein metabolism is a historically situated relation between biological bodies, ecological systems, and means of producing food. It is also a potent metaphor for understanding changes in these relations over time. As industrialized capitalist agriculture has had to confront the excesses, inefficiencies, and harm brought about by its practices, it has resolved to set about reorganizing nature in shifting configurations that implicate, and are in key respects entwined with, the contemporary life sciences.[32] The point we want to emphasize is that these reconfigurations of capitalism-in-nature include and require *the nature of the body itself*.[33] Active bodily practices such as the consumption of protein as a dietary staple, and the forms of athletic labor prescribed to metabolize those nutrients, are a crucial yet often overlooked aspect of such "organized revolutions" of capitalism-in-nature. Metabolic rifts and shifts are about grand world-historical processes reorganizing the world in the interests of their persistence, profiteering, and power. But the localized sites of these processes are bodily and mundane, as everyday as sipping on a postworkout smoothie or munching on a protein bar between meetings.

Once we move beyond the commonsense notion that demand for protein is primarily a cultural phenomenon driven by health and fitness imperatives and met dutifully by a confluence of technoscience and the profit imperative, the body itself is reconfigured in the process. It is no longer an "end user" of protein, no longer a singular entity, no longer easily distinguished from other animal bodies enlisted in the mass production and consumption of whey protein such as the cows whose incarceration is in service of reproducing dairy products. Instead, the metabolic capacities of human bodies can be seen as materially entangled in wider metabolic flows within and between capital accumulation and agricultural production, that take place within historically specific socio-ecological systems and relations. The desire for protein-enriched bodies within cultures that valorize working out or whey protein as a dietary staple plays a crucial, reiterative part in these circuitous processes. We can be more specific: The metabolic functions that inhere within human biology are *put to work* by the interests of capital, in this case the dairy industry, which are allied to and promulgated by technoscientific knowledge and innovation about the capacities of the body. The body emerges simultaneously as a biological agent with remarkable metabolizing qualities, animated by desires for vital, muscular, healthy bodies, and as an invaluable "fix" to the crises of the dairy industry. Metabolizing bodies are thus recruited not only as a fix to "disappear" the waste of milk and cheese production but also to create new avenues for accumulation and new markets for whey effluvium rematerialized as protein powder. Figured this way, it is no exaggeration to say that the bodies of many health and fitness enthusiasts have been enlisted to invisibilize dairy waste, to "work out" the excesses of dairy industrialization and make "whey the pollutant" disappear.

But no fix is absolute, whether spatial, technological, or, as in this case, socio-ecological.[34] While on one level we can say that the bodies of many health and fitness enthusiasts have been mobilized to process this surplus substance, this is not the whole story. Told this way, it risks giving the impression of a totalizing process, of capitalism as an inescapable, always efficacious historical force. What it misses is what Moore might call the metabolic shift: the deferrals and displacements that render its disappearance incomplete and fraught with its own problems.

Protein as a Pollutant

It is whey's nitrogen density that makes it such an unruly, pungent, and polluting force, more potent than raw human sewage by some magnitude.[35] This notion of protein as a pollutant is itself a strange proposition, akin to conceiving it as waste. Where protein has recently been cited as a pollutant, this most often owes to its association with methane. The 2014 documentary *Cowspiracy* brought the gaseous consequences of the industrialized rearing and slaughter of cows to a mass audience already sensitized to messaging around the deleterious environmental effects of not only meat consumption but consumption writ large. Eating meat, already cast as environmentally harmful and ethically compromised, was joined up explicitly with industrial agriculture and the problem of methane gas. Criticized in some quarters for its heavy-handed messaging and occasional privileging of ethical imperatives over scientific accuracy, *Cowspiracy* nonetheless played its part in articulating the consumption of meat to environmental harm and environmental politics. It also reaffirmed a long-standing discursive association between meat and masculinity, something Carol J. Adams has been alert to since at least 1991, when her *Sexual Politics of Meat* was first published.[36]

That methane emissions are ecologically hazardous is beyond scientific reproach. Noteworthy for us, though, is that protein-as-meat is not exactly the offending category here, not itself the pollutant under scrutiny. Rather, the methane gas produced by cows in industrialized agricultural settings is the *outcome* of their mass incarceration and slaughter, of the steroids they are fed and the numbers in which they are reared, of the production of demand for meat globally. It is demand for protein, cast as the cow's dismembered carcass or the dairy products it coproduces under these conditions, that implicates protein with environmental pollution.

Far less attention focuses on *protein as itself a pollutant* owing to its nitrogen density, as inhering a toxicity that is also key to its nutritional value. Urea excretion is the body's means of expelling some 80 percent of waste nitrogen, something it needs to do because we simply cannot metabolize all of the protein we consume, no matter how much athletic labor we endure. According to Stuart Phillips, an authority on protein in nutritional and exercise science, there is indeed a limit to the amount of protein the body can process. The specifics of that limit are as contingent and fluid as the other elements of protein metabolism discussed earlier in this chapter, but the reason for this deserves emphasis. Like other biophysical and ecological systems put to work by industrial capitalism, the human body cannot bear the burden of this nitrogenous intake entirely, and a significant amount or "large fraction" remains in the body postconsumption.[37] Once protein has been metabolized as muscle or fat, or excreted as waste, it remains as a "fundamentally toxic" compound.[38]

Even though humans are inefficient processors of nitrogen, this toxicity is not primarily a threat to the human body. The broader concern is anthropogenic nitrogen pollution, a pressing environmental problem that garners little public attention proportionate to its devastating effects, certainly compared to climate change, which has become almost a synonym for environmental ills writ large. According to Jan Erisman, a leading researcher in the science of sustainable agriculture, nitrogen pollution is "one of the most pressing environmental issues that we face," though it receives little attention compared to carbon and methane in particular.[39] It is worth quoting Erisman in full:

> Numerous, often interlinked, thresholds for human and ecosystem health have been exceeded due to excess nitrogen pollution, including thresholds for drinking water quality (due to nitrates) and air quality (smog, particulate matter, ground-level ozone). Eutrophication of freshwater and coastal ecosystems (dead zones), climate change and stratospheric ozone depletion are also consequences of the human modified N, cycle. Each of these environmental effects can be magnified by a "nitrogen cascade" whereby a single atom of reactive protein can trigger a sequence of negative environmental impacts through time and space.[40]

It is important to note that human and animal excretion is just one of a multitude of sources of anthropogenic nitrogen pollution. Synthetic nitrogen fertilizer, which acidifies soils and leaches into drinking water, rivers, and seas, is the major source of reactive nitrogen in the environment. Its global emergence has recently been charted as part of British imperial expansion and governance in the nineteenth century, thereby complicating histories that would reduce the proliferation of nitrogen fertilizers "to a simple story of triumphant modernity" or economic globalization.[41] Clear parallels with the emergence of whey protein can be drawn, not least in the tendency of supporters and detractors to hail protein as simply pollutant or health elixir, as problem or solution. While the world's population could not be fed without the use of industrially produced fertilizer, the demand for protein-rich foods has dramatically exacerbated nitrogen pollution. Even before humans consume a protein shake, then, substantial volumes of nitrogen are lost to the environment through the cultivation of feed for the cattle who produce the milk, and even more are lost in the cows' manure. And once excreted, the problem is compounded. This is because nitrogen is "recalcitrant" in nature, and its removal from wastewater is costly and energy intensive.[42] Multiple methods of urea removal are in development, but in the United States, for instance, only 5 percent of this toxic substance is currently removed.[43]

When we describe whey's manifestation as a toxin, as we have at various points through this chapter, we are pointing to its high nitrogen density and the implications this can have for the bodies of flesh and water that it comes to inhabit. We do not mean to convey that whey's commodification and ongoing toxicity undermine some purified notion of bodily innocence and integrity. Embodying toxicity is increasingly understood as simply part of living in the twenty-first century, when blood, cells, urine, and other vital fluids and tissues are shown to be "contaminated" by chemicals that inhabit the water we drink, the food we eat, and the air we breathe.[44] Rather than reifying whey's recalcitrance against an elusive notion of bodily purity,[45] its toxic properties are significant for their persistence in the face of commodification and technoscientific refinement, even once processed into powder, digested, or excreted. Owing to these unruly qualities, despite a thirty-year effort on the part of the dairy industry to tame its noxious waste by enlisting humans and animals in its regeneration, whey's toxicity persists in the same waterways it occupied in midcentury North America. And in this process of diverting excess whey into bodies and converting it into powder for consumption, waste becomes food, food becomes waste, and protein rematerializes anew.

Conclusion

Following whey protein's transformational journey through twentieth-century fluctuations in milk and cheese production reveals a powerful, pungent agent: a substance that does not simply disappear into the atmosphere on account of its constitution as waste or after its commodification. Understanding whey as a multiplicitous substance through the entangled processes of commodification, multispecies labor and the global nitrogen cycle demonstrates that ecological issues have not been allayed since the making of a protein powder industry created new possibilities for its use and profitability. Whey dumping and its known ecological effects endure: Only a fraction of whey is converted into commodity form for human consumption, and even this, in turn, becomes the problem of wastewater management engineers faced with filtrating increasingly nitrogen-dense water systems. The rest of the excess is either diverted back into animal feed or subject to costly ultrafiltration practices before finding its way, eventually, into those same wastewater management stations.

When we say that whey is active, this is not a granting of agency per se but a recognition that its propensies have been unleashed and amplified by the scale of industrial agriculture, which in turn has displaced and dispersed whey into circulations far beyond the farm. There is a hubris in the presumption that anything else would be true: that whey would simply accede to the will of capital and technoscientific mastery, as well as a shallow appreciation of how capitalism works and, more pointedly, does not work. Rather than entirely jettison whey's official biography, our account thereby troubles its claims to have left behind whey's history, not least its polluting capacities as a nitrogen-dense substance. In attending to this lethal potency, which is also the source of its nutritional qualities, whey teaches us something important about its commodification, namely that it is the *socio-ecological relations* between its capacities and the imperative to profit from them that go toward explaining its extraordinary contemporary status as a ready-to-mix übernutrient. Whey remains in this sense as "baffling" and "unruly" as protein has for some two centuries, continually refusing to acquiesce entirely to attempts to delimit its boundaries and to profit from its capacities.

As such, this chapter recasts whey's role in its own commodification. There is no denying that the confluent interests of technoscientific food engineering and capital's relentless need to renew itself combined to make whey palatable and, through a discursive apparatus of marketing and scientific endorsement, desirable. It's just not the whole story. After all, had whey succumbed to its intended end as waste and settled silently on riverbeds, a dutiful servant in the mission of making cheese in abundance, then the "problem" of whey waste might never have been transmogrified into the "solution" of whey powder. This in turn would not have happened if not for the extraordinary metabolic capacities of human (and other animal) bodies, each ingesting and processing protein powder in its myriad forms, breaking down whey's dense properties through different kinds of labor and excreting what remains in protein-rich urea. What is the body conceived this way? An experimental site? A biochemical processor? A vector for affective attachments? These questions stay in play as we continue to follow protein out of the world of industrialized dairy and into the cultural milieus of protein consumption.

4

A POVERTY OF FLESH?

Sarcopenia, Aging, and the
Economization of Protein Deficiency

We physicians all know about renal insufficiency and heart failure and respiratory failure, but we'd never thought about muscle failure.—Alfonso Cruz-Jentoft, quoted in Drew, "Lifting the Burden of Old Age" (2018)

Following the commodification of whey waste as a supplement for mass consumption in the 1980s and 1990s, the need to find homes for this newly abundant protein source only expanded. Milk-based powders and supplements had been used in niche bodybuilding subcultures and hospital settings since the early twentieth century, but these modest outlets were simply not proportionate to the supply made possible by whey's socio-ecological transformation from gutter to gold.[1] New markets for protein, tethered to new and enduring bodily anxieties and desires, were needed. And while the persistent association between protein and muscle growth provided health and fitness proselytizers and sport and exercise scientists (groups that often overlap) with a ready-made route to pursue in this endeavor, the potential of protein as way to treat *muscle loss* also emerged as a market in the making.

Sarcopenia, broadly defined as the loss of muscle mass and function over time, was "discovered" in the late 1980s. That discovery, and the subsequent adhering of muscle loss to notions of (un)healthy aging, the "global challenge" of population growth, and the pursuit of protein as a solution, are the focus of this chapter. Hundreds of scientific articles, reviews, and reports have since explored sarcopenia's causes, symptoms, and treatments, culminating in the World Health Organization classifying it as a disease in 2016. Despite this outpouring of scholarship and recognition, knowledge about sarcopenia is haunted by the notion that losing muscle and strength are simply signs that a person is advancing in years. The main factor that puts one at risk of sarcopenia is entering older age, though as with the other known factors—such as physical inactivity and osteoarthritis—it is difficult to prize apart the causes from the symptoms. While it is variously described as a condition, a disease, and even a disability, sarcopenia is difficult to decouple from the already nebulous category of aging itself. This raises broader questions about how and why sarcopenia has come to occupy such a prominent place in contemporary health, exercise, and nutrition sciences and their allied industries: How did aging become a medical problem? How might we explain why a decline in muscle mass became a condition of scientific concern, and later a classified disease? And what is the role of protein in adjoining sarcopenia to consternation about population growth and animating the compulsion to intervene in our bodies under the auspices of self-responsibility and "healthy aging"?

In response to these questions, we explore the emergence of sarcopenia as a condition of age-related muscle loss, to which dietary programs of protein ingestion combined with age-appropriate exercise are among the leading proposed remedies. Instead of proceeding from a notion of sarcopenia as a timeless and self-evident biological phenomenon only recently known to science, we situate its emergence at a moment in which the notion of individual culpability for aging became conceivable, diagnosable, and intervenable. This is a moment marked by the political and economic ascendance of neoliberalism, a formation characterized by privatization, structural adjustment and trade policies, austerity programs, a fetishizing of market logics as indicators of social value, aversions to state spending on healthcare and other public services, a championing of individual responsibility, and exacerbated inequality. We argue that sarcopenia's emergence in the late 1980s is best comprehended in this conjuncture of forces, and we suggest that the rise of individualized "healthy aging" imperatives and widespread fears about the cost of an aging population have coalesced in the creation of a novel frontier for protein researchers and marketers seeking new homes for protein knowledge and products.

Through sarcopenia, protein has been adhered to aging, extending its established homology with muscle growth to become a leading solution to the muscle loss that comes with advancing years. As the subject of numerous studies to determine the type, amount, and duration of intake required to manage sarcopenia, and the focal point of a burgeoning market in nutritional supplements geared toward older adults, protein has emerged at the center of a growing lifestyle-knowledge economy centered on "healthy aging." Consumer research suggests that the value of the market for senior nutrition stood at just over US$23 million in 2022, with protein representing 41.9 percent of the demand enumerated by nutrient type, more than double its nearest rival—iron—for market share.[2] Protein according to industry news source *Nutrition Insight* is "the name of the game" in shaping consumer demand and driving product development in the healthy aging industry.[3]

In addition to its role in the circulation of healthy aging imperatives, we have found that contemporary sarcopenia research mobilizes the widespread discourse through which demographic aging, or more accurately the shift in the distribution of some groups of people in some parts of the world toward older ages, has become recognized as a global challenge. The fact that COVID-19 has "triggered an unprecedented rise in mortality" that has "translated into life expectancy losses around the world" does not seem to have tempered the discourse on population aging, with the United Nations warning in 2023 that "population aging is poised to become one of the most significant social transformations of the twenty-first century, with implications for nearly all sectors of society, including labor and financial markets, the demand for goods and services, such as housing, transportation and social protection, as well as family structures and intergenerational ties."[4] These same implications are invoked in sarcopenia research as a rationale for the stakes involved in its onset, a formulation in which aging as a phase of life becomes a site of capital accumulation and biopolitical interest. This, we venture, is the simultaneous economization and medicalization of aging, whereby demographic aging emerges as a global challenge that can be both monetized and made subject to clinical intervention, while the cost-of-living crisis, the demise of the welfare state (in places where it actually existed), deferrals of pensionable age, social isolation, and other systemic factors that make aging difficult proceed apace.

As we illustrate in what follows, economization and medicalization manifest in two key ways: In the costing of an aging population in the overlapping realms of scientific, state, and corporate research, and in the creation of "interventions" through which individuals can offset muscular deterioration and the manifold problems to which it is affixed (which are of course vast, given that sarcopenia is at times near indistinguishable from the nebulous category of aging itself). Through these techniques, and despite this ambiguity, sarcopenia has become a portent of advancing years on earth with consequences for how those years are then experienced. Muscle loss in aging is understood as inexorable and exorbitantly expensive for individuals, healthcare providers, governments, and society writ large—*unless* one employs certain practices of the body in mitigation. In this sense, sarcopenia is inevitable but also pliable, and to a point even reversible. As one of two primary forms of mitigation proposed by experts, protein is inextricable from sarcopenia. Following protein through the discourse on muscle loss in older age helps illuminate the economic and social underpinnings of sarcopenia and of healthy aging more broadly. Indeed, protein is crucial to what makes sarcopenia simultaneously costly and profitable, rendering old age a process of deterioration and accumulation. In following protein into this realm, and disentangling this knot of cultural, scientific, political, and economic forces, we contend that sarcopenia is "made up," to use Ian Hacking's phrase.[5] Constituted variously as a disease, a condition, or a disability afflicting those of advanced age, sarcopenia generates a prime population for intervention by exercise and nutrition scientists and protein prospectors alike.

Sarcopenia, or Aging as a Disease

At a New Mexico symposium in 1988, Professor of Nutrition and Medicine Irwin Rosenberg took the established tie between muscle, health, and vitality and dramatically expanded its horizons. Addressing colleagues well versed in the medicalization of aging and so presumably amenable to interventions in the aging process, Rosenberg proposed that the loss of muscle mass and function as a person grows older was seriously understudied and should therefore be focalized in future research and clinical practice. Since the symposium had been convened to discuss the "health and nutrition of elderly populations," his proposal met with a receptive audience.[6] Lay folks might have found it lacking in originality, though. After all, that muscle mass usually decreases as people get older is hardly a novel observation. Surely it is just a sign that one is advancing in years and so really just a proxy for aging itself?

That New Mexico symposium is now recognized as the birthplace of sarcopenia, the word Rosenberg coined that day to describe muscle loss as a condition, or disability, afflicting older adults. It was also where the corollary was advanced that this condition is wrapped up with other physical, social, and financial problems of aging, thus demanding intervention from, and investment in, nutritional, health, and exercise science. Allied with resistance exercise, dietary protein has been a key component in treatments for sarcopenia ever since.

Sarcopenia is etymologically rooted in Greek and means "loss or poverty of flesh." From this initial act of inscription, Rosenberg showed a wry awareness of what invoking an ancient lineage could do for popularizing muscle loss as a medical condition. When a word is rooted in Greek, it takes on a kind of proleptic aura, as if its etymological inheritance authenticates the object's own timelessness and affirms its enduring import. Think of it as a technique of authentication that instills the newly named object with a venerable air, of not only being an observable biological reality but also sharing a genealogy with the wisdom of the ancients. It also conjures an imagining of ancient Greece as the birthplace of philosophy, science, and civilization, and Europe as the origin of the modern world. Just as the word *protein* was taken from *proteios* to connote its primacy, so the naming of sarcopenia exhibits this validating effect.

The name, in other words, was neither accidental nor incidental to the subsequent growth of sarcopenia the medical condition. Rosenberg later reemphasized that a formal name akin to *osteoporosis* or *osteopenia* was always intended to realize particular objectives, for "the issue of what will be taken seriously by the research community, the practicing community, and by funding sources is more than a trivial matter."[7] At the same time, he made efforts to circumscribe his own role in merely naming and drawing attention to a phenomenon that long preceded him, writing that "the history of sarcopenia is as old as the aging of man, even if the term sarcopenia is but 20 years old."[8] All he was doing was attaching a label to a preexisting phenomenon, a label that was suitably venerable for the origins of the condition and the intended effects of naming it.

This framing of sarcopenia as a timeless phenomenon only recently recognized as a scientific object begets deeper philosophical questions about names and referents, words and things. There doubtless were people whose muscle mass and function decreased as their years on earth advanced long before the term *sarcopenia* entered the scientific lexicon. We can say this not because we know it as a transhistorical cross-cultural truth with empirically verifiable records, but because it makes sense to speak of the "primordial materiality" of human bodies in this way, as finite organisms that live, decline, and die.[9] This much is uncontroversial. But to invoke sarcopenia as a condition or disease is to take part in the medicalization of senescence, whereby muscular deterioration is drawn out from the realms of the inevitable fate of all who survive into later life and becomes a kind of red zone for health-conscious seniors, a site of untapped potential to be diagnosed, anticipated, and subject to intervention.[10]

Sarcopenia is now a recognized medical condition, as illustrated by its aforementioned entry into the World Health Organization's classification of diseases in 2016. Yet we might also understand it as a symptom of the medicalization of aging, an effect of the figuring of older age as a multifaceted problem. The notion of medicalization was already influential by the time of Rosenberg's proclamation. How exactly this process manifested itself had been fiercely debated through the 1970s and 1980s, with leading figures emphasizing either power or progress: the extent to which medical knowledge and authority was a form of social control or imperialism, or an advancement that improved quality of life. Medicalization has since become so prevalent as to provide enough evidence to attest to each of these positions and more, and in this sense the claim that sarcopenia represents the further "medicalization of aging" is unremarkable, somewhat vague, and in ample company. This ubiquity goes some way to explaining why it has not been explored in the enormous literature on medicalization that has now transcended the sociology of medicine to take hold in the wider social scientific conceptual repertoire, where examples of medicine's encroachment into so much of everyday life abound. In any case, for those who have entered the "aging population" in recent decades, for whom the experience of getting older is already medicalized in multiple ways, sarcopenia represents another diagnostic means of understanding changes to the body over time.

Sarcopenia conforms and contributes to multiple theses on the medicalization of aging, including the presumption of frailty in older people that follows from its major premise: the "age increase–muscle decline" correlation. The tethering of muscle loss to aging is present across sarcopenia research, and getting older is the condition's most recurrently posited overarching cause and its bodily manifestation. In 1995, William Evans, a collaborator of Rosenberg's, defined *sarcopenia* in the *Journal of Gerontology* as the "age-related loss in skeletal muscle mass."[11] Some have since distinguished loss of muscle strength from loss of muscle mass, preferring the term *dynapenia* to describe the former.[12] But the majority of studies continue to employ Rosenberg's general definition and figure muscular loss, strength, and function together as sarcopenic symptoms. Reflecting on his initial remarks in 1988, Rosenberg wrote the following in 1997: "I noted then that no decline with age is as dramatic or potentially more significant than the decline in lean body mass. In fact, there may be no single feature of age-related decline more striking than the decline in lean body mass in affecting ambulation, mobility, energy intake, overall nutrient intake and status, independence, and breathing."[13] Striking indeed. Here Rosenberg declares a litany of vital functions—moving, breathing, eating—to be correlated with an age-related decline in muscle mass, making muscles and their failure over time a problem of far-reaching consequence for all who live long enough to experience it. From the outset, aging and muscle loss are tethered not only to each other but also to these fundamentals of bodily existence.

In this appeal to sarcopenia's reach and relevance, Rosenberg unwittingly signposts its enigmatic qualities: It is a condition that is said to compromise the most fundamental dimensions of bodily existence, yet it has proved difficult to distinguish from the senescence of cells that accompanies and characterizes aging, and is therefore difficult to delimit at all. Writing in *Nature* on sarcopenia and the "burden of old age" in 2018, writer and biologist Liam Drew stated plainly the dilemma facing sarcopenia research: "One of the main difficulties in defining sarcopenia as a disease is that a degree of deterioration of the body has been taken as part and parcel of getting old for millennia. Muscle begins to form in utero, and then grows until it reaches a peak mass, usually in a person's late 20s. From then on, there is continual loss."[14] That question once again arises: If this loss is continual and generally experienced by those who live into their thirties and beyond, then how to distinguish the afflicted from those who are simply aging? Echoing this ambivalence, reviews of the subject include a range of caveats, noting that "sarcopenia remains a condition with neither an agreed on definition nor an effective treatment,"[15] that "much like the underlying causes of aging, the biology of sarcopenia remains elusive,"[16] and that despite its newfound status as a disease, "a diagnosis of sarcopenia is very rarely made or documented in medical records."[17] Definitional difficulties concerning its etiological origins and causes persist today, making the correlation of aging and muscle deterioration at once sarcopenia's key premise and the source of its greatest uncertainty.

Defined parameters notwithstanding, the relationship between muscle mass and the rate at which it declines with age remains as close as sarcopenia research gets to meaningful consensus. Yet even within this relationship, difficulties in measurement arise. For example, Taylor Marcell offers the analogy of two contrasting physiques to illustrate the limits of calculation in this muscle-longevity equation, asking, "Who really is sarcopenic?"

If we compare the muscle mass of body builder turned actor/gubernatorial candidate Arnold Schwarzenegger to that of the 1984 Olympic gold medalist in the 1500-meter run, Sebastian Coe, who is sarcopenic? A 70 percent reduction in mass is suggested to lead to disability, yet Arnold could afford to lose 70 percent of his lean body mass and still have greater muscle mass than that of Sebastian. Thus, if we assume that muscle loss rates are similar, then the greater the (starting) reserve capacity the longer it will be before sarcopenia or physical frailty will compromise function.[18]

Readers could be forgiven for inferring from this analogy that the secret to avoiding or at least offsetting sarcopenia lies in increasing one's "reserve capacity" of muscle to such a degree that loss will not be debilitating in later life—in which case, Schwarzenegger becomes an icon for longevity, and muscle mass something that one should build and "store," much like wealth, in mitigation of anticipated loss. But how exactly does one do this, given that muscle is not stored but constantly maintained and rebuilt, as we are all reminded constantly in mantras such as "Use it or lose it"? In any case, are we really to embrace the claim that Schwarzenegger's physique should be emulated to stave off the "disease" of muscle loss in later life? Muscle mass as an indicator of and insulator against sarcopenia therefore serves to confound matters rather than clarify them, as Marcell's example attests.

Agreement on what exactly constitutes sarcopenia and how and why it emerges eludes even those who make this question the central object of their scholarship. In 2009 a European Working Group on Sarcopenia in Older People (EWGSOP) was formed with the goal of defining sarcopenia for clinical applications.[19] The EWGSOP's first guidelines were published in 2010 and were swiftly followed by guidelines from the US-based International Working Group on Sarcopenia in 2011, and the Asian Working Group for Sarcopenia in 2014.[20] Read these reports and you will be presented with algorithms, mechanisms, tests, tools, metrics, and other representations intended to decipher and demonstrate sarcopenia's inner workings. Some posit that sarcopenia has a "neurogenic origin" that will "escalate the disease progression";[21] others link it to insulin, testosterone, and other hormones. All probe the inner workings of the body in pursuit of sarcopenia's elusive etiology. Similar uncertainty surrounds the relationship, if any exists, between a diagnosis of sarcopenia and improved outcomes. As Christoffer Bjerre Haase, John Brodersen, and Jacob Bülow state in a 2022 issue of the *British Medical Journal*, "There is currently no evidence to support that a diagnosis of sarcopenia improves prognosis. No evidence shows any differences between sarcopenia treatment and general recommendations for physical exercise and diet. Moreover, the current diagnostic cut-off points, including gender and regional specifications, are arbitrary and non-validated. Evidence to distinguish between normal and pathological age-related loss of muscle mass does not exist."[22]

This elusiveness is in itself unremarkable. Many objects of scientific inquiry, including matters of human biology, do not have absolute consensus around their origins or effects. Even death, that most decisive of biological conditions that concludes the aging process, is surprisingly complex and difficult to delimit—more so, paradoxically, for those who study it most intimately.[23] This is not simply because death has culturally diverse meanings but also because the physiological process and pronouncement of death have proved confounding to scientists and medical professionals just as they have in spiritual and religious traditions. All this is to say that holding sarcopenia to trial on account of its facticity is not to denounce it entirely, as if certainty were the only goal. Those in the field of social gerontology have made plain that aging is not a linear process, nor is it experienced uniformly. Physical decline with age is not simply fated. How we live our lives matters, as does where we live, and what kinds of duress and care we experience along the way. In other words, the end of biological life is inevitable, but how it unfolds is patterned unevenly by the stratifications of social life, not least by financial security, social integration, a conducive built environment, proximity to fresh foods, and access to essential services. This context is lost, or perhaps never grasped, when aging is defined as the wasting or senescence of cells. When these uncertainties go untroubled, physical decline becomes a naturalized consequence of aging, and aging itself becomes a problem, in this case a disease requiring clinical intervention. This is aging as a limit to treat or transcend, rather than a phase of life. An alternative approach, and the one we prefer, is to consider sarcopenia as the product of a particular historical conjuncture. Here we widen our analytic lens to consider why the late 1980s was such a fertile moment for the notion of muscle loss in later life to emerge as a problem that could and should be addressed.

Neoliberalism, Healthy Aging, and the Body Intervenable

Since the 1970s, fitness in the industrialized West has taken on unprecedented moral and material significance. During this time, mass participation in exercise has exploded, placing fitness at the very center of contemporary ethical subjectivity and the marketplace for self-fashioning.[24] In North America, the upsurge of mass participation in exercise became known as the "fitness boom," a term denoting not just the increased popularity of physical activity but also what Jürgen Martschukat, following Michel Foucault, calls the fitness "dispositif," or apparatus—"an era-defining network of discourses and practices, institutions and things, buildings and infrastructure, administrative measures, political programs and much more besides."[25]

It was during this period that the social import of fitness became inextricably linked to the social import of individual responsibility for health. While fitness operates in everyday language as a synonym for the vigor and well-being that can derive from regular participation in exercise, its popularization has prompted kinesiologists and other scientists with professional stakes in its meaning to generate more technical definitions that tie it directly to measurable, medicalized indicators such as body composition, cardiorespiratory endurance, muscular strength, muscular endurance, flexibility, balance, and speed of movement, often with an eye to the vaunted twin poles of elite sports performance, on the one hand, and illness prevention, on the other.[26] In these overlapping, mutually reinforcing discourses, fitness emerged as the route to health, a concept that became its own weighty dispositif in the political order of the 1980s—what Robert Crawford calls a "super-value" or a "metaphor for all that is good in life."[27]

The nexus adjoining health, the individual pursuit of fitness, and the political and economic doctrine of neoliberalism was forged in this moment, when the "fit," "healthy" body became neoliberalism's biophysical signifier and constituent symbol. The power of these connections is to render each part assimilable and beyond reproach, such that it is virtually impossible in today's world, as Jonathan Metzl and Anna Kirkland have contended, to stand "against health."[28] They write, "Health is a term replete with value judgements, hierarchies and blind assumptions that speak as much about power and privilege as they do about well-being. Health is a desired state, but it is also a prescribed state and an ideological position."[29] As such, health, and cultural investments in so-called healthy lifestyles have come to prop up a system of economic relations that depends on what Crawford calls "healthism" for its reproduction.[30] Crucially, given our focus in this chapter, healthism situates the problem of disease, and solutions for preventing and treating it, firmly at the level of the individual and squarely in the realm of the market. Indeed, under neoliberalism, "personal responsibility for health is widely considered the *sine qua non* of individual autonomy and good citizenship."[31] In this regime, subjects have come to "define themselves in part by how well they succeed or fail in adopting healthy practices and by the qualities of character or personality believed to support healthy behaviors."[32] Although the ability of individuals to act on health discourses is acutely shaped by their access to material and cultural capital, such constraints have become largely invisible under the normalizing logic of health as a super-value. As such, ill health has come to be seen as a reflection of poor "lifestyle choices" and inadequate commitment to bodily well-being, even as the broader effects of neoliberal policies have been repeatedly shown to exacerbate health inequalities.[33]

Crawford further suggests that the "new health consciousness" that gained ascendancy in the latter decades of the twentieth century did not simply reflect the emergence of neoliberalism but in fact played a decisive role in the formation of this mode of organizing political and economic life. A widespread awakening to environmental dangers and the effects of smoking were particularly important in producing concerns about "lifestyle" hazards, an interest in changing what came to be understood as risky behaviors, and a new "take charge" discourse of personal responsibility for health.[34] Although corporations and government officials with a political and economic stake in a turn away from collectivism and regulation mobilized ideologies of individual responsibility for living healthily and staying in shape, these ideologies grew alongside white middle- and upper-class investments in the body as a route to personal renewal and a secure future. While other factors (such as the decline of organized labor) intersected with the ascendance of personal responsibility over collective care, Crawford suggests that the notion of individual liability for health "played a decisive role" because the preoccupation with the healthy body became a model for what individual responsibility or lack thereof would differentially grant. Health consciousness thus came to operate as "an embodied replication of individual responsibility for economic well-being" and opened the way for attacks on the welfare state (such as it was in the context of the United States, Crawford's primary focus) and on collective and relational modes of living writ large.[35] "What has become clear in hindsight," writes Crawford, "is that individual responsibility for health, although not without challenge, proved to be particularly effective in establishing the 'common sense' of neoliberalism's essential tenets."[36] In this context, it is no surprise that the mass turn to "working out," and the concomitant appearance of numerous new fitness-related commodities, became defining features of the rise of health as a super-value.

More than four decades since Crawford's incisive critique of healthism as a buttress to the ascendance of neoliberalism, we can see how disease causation, illness prevention, and wellness strategies designed to optimize the body's capacities and secure its "best possible future" have become central points of common interest and near-axiomatic cultural values.[37] Health and fitness are now the basis of powerful industries and consumer cultures, the subject of complex forms of lay knowledge, and key to the discourses through which contemporary identities are shaped and performed, especially in class strata where access to resources helps mitigate the iniquitous effects of neoliberal capitalism. And while the face of these industries and cultures is often youthful and vigorous, responsibility for "healthy aging" extends across the lifespan for those living in societies shaped by neoliberal rationalities and healthism, societies where those same rationalities have drastically eroded collective healthcare provisions.

Sarcopenia is born out of this neoliberal formation of heightened health consciousness and its cultural homology with the pursuit of fitness as a mode of responsible selfhood. Its key premise of muscle decline with age figured as a medical problem animates "successful aging" discourses that establish the self-governance of a person's own aging body as an act of good neoliberal citizenship, where the "risks" inherent to aging are assumed by individuals—or, rather, those with the resources to meet its challenges. These discourses also simultaneously reinforce and unsettle dominant notions of frailty in older age, for sarcopenia is at once an indicator of impending frailty *and* a key rationale for the activities of the health and exercise sciences and their adjacent industries. The aging body here is configured as an inevitable problem and source of financial, social, and bodily risk that thus requires investment in antiaging measures. Elaborating a similar argument as part of a polemic against the reach of capitalist logics and values into all kinds and all phases of human life, Jonathan Crary writes,

> To suppress aging is to imagine life as a distended present, suspended from time and shielded from decay or change. For thousands of years, the finiteness of life has been what has given meaning, passion, and purpose to our existence, and to the ways we love and depend on others. To debase human finitude by proposing that individual longevity could become a sought-after biotech product for the affluent is part of the extinguishing of any values or beliefs that transcend the voraciousness of capitalism.[38]

Targeting aging as a limit to be deferred or even defeated is part of a transhumanist, promethean fantasy that finds expression in a dream of immortality, one that Crary points out is increasingly pursued by tech-obsessed billionaires. He continues,

The commodification and privatization of the future is now explicit, as "time to live" is assimilated into the logic of financialization. An anti-aging industry incites anxiety and fear—fear of frailty and dependence in a world in which most social forms of support have been weakened or eliminated. Even under minimal welfare state provisions, old age was a structural problem for capitalism because of its relative unproductivity and diminished consumerism. Now, "aging" becomes part of the current precariousness and disposability of all human life.[39]

Crary's commentary lends itself to an understanding of sarcopenia as capitalizing on a fear of physical frailty that is made more real by the structural weakening of state healthcare provisions. But there is a paradox at work here that demands a reckoning: Greater life expectancy is one of the health indicators that is seen in countries where neoliberal policies originated, notwithstanding the damage of "shock therapy" economics, growing inequality, and global pandemics.[40] Without yet delving into the stratifications within these general trends, the rise of neoliberalism has, at least until 2020, been accompanied by greater overall longevity. Is this a paradox, a classic contradiction of capitalism? If we confront the aging population not as a limit to accumulation but as a growth frontier in its own right, then perhaps not.

In the scholarly and commercial interest marshalled around it, sarcopenia does not only or simply represent a "structural problem" for capital accumulation, even though the condition is generally portended as a limit to a person's active participation in society. While old age is often treated in this way in industrial capitalist societies that organize groups of people and phases of life based on their presumed ability to perform wage labor, sarcopenia is better understood as part of the establishment of older age and "later life" as frontiers for capital accumulation. This makes sense in the context of the fitness boom and participation in physical activity as a key signifier of successful aging, as Crawford predicted in his analysis of healthism. It makes sense once we situate sarcopenia as emerging concomitantly with a political and economic doctrine that seeks to reduce public spending on healthcare, and eldercare especially, while fetishizing market logics as the fundamental and indivisible indicators of social value. And it makes sense that under these conditions, protein deficiency and muscle loss became new frontiers for protein marketers and researchers seeking new homes for protein knowledge and products. From here we can begin to venture a definition of sarcopenia that retains its basic premise, its tethering to age, but sees its medicalization as inseparable from its economization.

Economizing the Aging Population

Contemporary debates about demographic aging are constitutive of broader anxieties around population size that usually unfurl with a stark warning about the sheer number of humans now inhabiting the planet. From a billion people at the end of the nineteenth century to nearly 7 billion a century later, that growth is commonly framed as an era-defining global challenge or even crisis. Projections for what this means for human life then veer off in different directions. Dominant among them is a neo-Malthusian camp, caught in the conservative consternations of an eighteenth-century pastor named Thomas Malthus, who predicted that unabated population growth would surpass food production capacities. Malthus was also concerned about the threat of popular rebellion against church and state arising from population growth, especially among the unruly or ignorant poor, and thereby cited it as something to be stifled and controlled for political as well as agricultural reasons. The spread of this logic to and from colonial, financial, eugenic, and environmental contexts has been well documented, and can still frequently be found in proclamations that the planet is overpopulated and that "excess people" are expensive and unsustainable.[41] Such discourses are often targeted at poor and racialized people and places, and at women, whose reproductive capacities are to blame. But increasingly the *age* of the population is part of this battleground in which who counts as "too many" is counted and costed as a financial threat to the future.

The configuration of an aging population as a key problem of the late twentieth and early twenty-first centuries is a revealing admission about how age has come to be governed and valued. The claim that there are "too many old people" is generally attributed to the baby boomers entering their seventies, their numbers owing to lower birth mortality rates and postwar economic and social policies, as well as declining birth rates among subsequent generations. The same claim is often racialized when referring to older white Westerners in rich countries, as opposed to young, brown people in the Global South, each figured as too numerous.

The implications of these demographic shifts are borne out through biopolitical analyses of national populations articulated in government reports, articles, and bulletins, each projecting demographic trends into financialized futures. A US Census Bureau article on "the graying of America" states that by 2035, "people age 65 and over are expected to number 77.0 million . . . while children under age 18 will number 76.5 million."[42] The report collates and analyzes death rates, infant mortality rates, migration, and other demographic factors to make these and other projections. It concludes that "by 2060, nearly one in four Americans will be 65 years and older, the number of 85-plus will triple, and the country will add a half million centenarians."[43] Many other countries employ this same calculative logic to their own national populous and had identified these demographic shifts before the United States did. For instance, a 2017 report by the Canadian Institute for Health Information led with the claim that "over the next 20 years Canada's seniors population is expected to grow by 68%," and the number of people aged over seventy-five is expected to double.[44] Japan has long been known for having the world's oldest population and has for some time been projecting a significant overall population decline. The United States might be late to seeing or identifying this trend, but its concerns about population aging are tantamount to those of other nations.

These projections perform not only a biopolitical reckoning with populations trends, something that Michel Foucault identifiedin in *The Birth of Biopolitics* (2004) as a defining mode of governance in modern nation-states, but also an economized understanding of longer life. They echo fiscal conservative concerns about public spending and debt that became widespread in the 1970s as part of the ascendant political economic arrangements of neoliberalism. Whereas M. Murphy's *Economization of Life* (2017) focalizes the costing of yet-to-be-born populations, marking out the female body as the site of intervention, here the logic of financialization is applied to an aging population and their cost at the end of life. It is a form of what Melinda Cooper calls "generational accounting," an economic equation that's inclined to frame public spending, debt, and Social Security as bad while threatening to incite and exacerbate intergenerational tensions. Cooper notes that the "baby boomer" generation were "identified as a distinct age cohort at the very moment the postwar welfare state was coming under intense attack," and were "from the beginning synonymous with fiscal indulgence."[45] That framing has continued through to representation of the cost of their care in later life as a burden on the future. US Commerce Secretary Gina Raimondo exemplified this sentiment in 2021 when stating that America's aging population would soon hit the country "like a ton of bricks" without the $400 billion in federal aid sought by President Joe Biden.[46]

Though headlines about eye-watering public costs are the most striking and visible expressions of generational accounting, how exactly the aging population is costed varies significantly. The discourse we have sketched above frames extended life as an imminent, expensive crisis, but there are cases where more balanced accounting prevails. In England, for example, a

2021 report by the endowment-funded Health Foundation charity subtitled "How Aging Affects Health and Care Need in England" appraises both the economic costs and benefits of an aging population. Rather than simply projecting existing healthcare costs in England onto future populations, the authors note a fall in the proportion of older people needing social care as the overall number of seniors has risen, while the need for NHS services for those aged over seventy-five with long-term health conditions has increased. As with other such forecasts, the authors find that the number of people living into their eighties is set to rise, but they caution policymakers that this does not mean all will require the level of care services that this age group has had in the recent past. Moreover, the report and others like it do acknowledge that the aging population make many economic contributions, albeit in roles that tend not to be counted or fully rewarded in the formal economy, such as volunteering and familial caring roles. There is an important recognition in this report, often absent elsewhere, that those people living longer than their ancestors are not just drains on an already strained welfare state. But in seeking to square these "benefits" of elder citizens with the financial "challenges" they pose to the United Kingdom's National Health Service, the economizing logic of generational accounting persists.

Differences aside, aging population reports share in evincing a wider political shift toward economic policies that advance the logic of prospecting and investment-reward calculations about those in later life, just as they do for those in all stages of life and for life in all its forms. In doing so, they form part of what Katherine E. Kenny calls "the reconceptualization of life as a revenue stream."[47] These policies serve to responsibilize individuals for the costs that tend to follow from getting older, further exhibiting the individualizing logic synonymous with neoliberalism. In the United Kingdom and elsewhere, this can be seen in policies to extend working life by increased retirement and pensionable age. Through the 2007 Pensions Act the State Pension terms were equalized for men and women, and the age at which the pension can be drawn was slated to rise from sixty-five to sixty-six in 2024 and to sixty-eight in 2044 "to reflect increasing longevity in society and make the State Pension affordable in the long term."[48] These changes form part of intergenerational tensions that continue to constitute the "baby boomers" as a coherent, uniform demographic group characterized by excess whose privileges have resulted in longer lives that now need paying for.

Sarcopenia has come into being at this conjuncture. It attracts a growing scientific mandate at the historical moment in which the aging population is the subject of state demographic projections, healthcare costing calculations, revised retirement and pension policies, and the rapacious commodification of all forms and phases of biological life. Ever since Rosenberg cited the appeal of the term *sarcopenia* to funding agencies when explaining why the condition would benefit from a name with ancient Greek lineage, sarcopenia research has joined in advancing an economized understanding of later life, primarily as a justification for continual research and investment. Sarcopenia researcher Taylor Marcell articulated this point as follows: "The sequela of sarcopenia may contribute to frailty, decreased capacity for independent living, and subsequent increased health care costs."[49] In 2021, Thomas Gustafsson and Brun Ulfhake made the same case: "The rapidly changing composition of the human population impact the incidence and the prevalence of aging-induced disorders such as sarcopenia and, henceforth, efforts to narrow the gap between health span and lifespan should have top priority.... The inflation of the aged over the younger cohort and the exponential growth of care expenditures, lead to an economic and social stress on society that needs to be addressed."[50] In these appeals to the "problem of the aging population," which are repeated across the sarcopenia literature, any biological determinism one risks in naturalizing conceptions of aging as simply cellular senescence fades into the background and is transposed by a patently socioeconomic rationale for why sarcopenia demands urgent attention. Most sarcopenia research recourses to aging population discourse in advancing the rationale that research and intervention in the loss of muscle mass and strength among older adults is a financial concern for both the individual and the state: As both individuals and national governments risk being burdened with the cost of aging, funding for research that mitigates these issues is imperative. The WHO's 2016 classification of sarcopenia as a disease takes on further significance here, as its ICD-MC Diagnosis Code "can be used to bill for care in some countries," cascading this financialized logic into an individual's dealings with health insurance companies and incentivizing physicians to identify muscle loss and treat it as disease.[51]

When sarcopenia research recourses to aging population debates and the burdensome costs that accompany them in order to demonstrate the need for ongoing scientific scrutiny, medicalization and economization are mobilized together as dual, entwined forces. A well-cited 2019 study by Scott Goates and colleagues titled "Economic Impact of Hospitalization in US Adults with Sarcopenia" presents an analysis of survey data on hospitalizations in the United States alongside population estimates from the US census. It proceeds on the assumption not only that sarcopenia is a disease but also that it is a disability or a disability-in-waiting, one with a considerable "economic burden" that the research simultaneously sets out to establish. The results take the form of a biopolitical analysis of an aging population stratified by age, sex, and race:

> The total estimated cost of hospitalizations in individuals with sarcopenia was USD $40.4 billion with an average per person cost of USD $260. Within this category, average per person cost was highest for Hispanic women (USD $548) and lowest for non-Hispanic black women (USD $25): average per person cost was higher for older adults (≥65 years) (USD $375) than younger adults (40–64 years) (USD $204) with sarcopenia. The total cost of hospitalizations in individuals with sarcopenia (≥65 years) was USD $19.12 billion. Individuals with sarcopenia had greater odds of hospitalization (OR, 1.95: $p < .001$) compared to those without and had an annual marginal increase in cost of USD $2,315.7 per person compared to individuals without sarcopenia.[52]

Here the aging population is disaggregated into categories of more or less costly social groups, albeit without a great deal of explanation. To what extent can these costs be attributed to sarcopenia, not least given that its status is still being debated by research communities? What are the implications of breaking these costs down into social groups that disappear as quickly as they appear? Many questions arise from this stratification, and they mostly go unanswered, but it is worth noting that the analysis fails to account for any social or economic value that individuals or groups may or may not add relative to these costs. Of course even stating this absence is to join in the biopolitical practice of generational accounting, rather than taking issue with the financialization of populations as an epistemic tool of neoliberal governance. Suffice it to say that the economization of sarcopenia easily slips into ossifying social categories, which emerge from their analysis as naturalized groups whose healthcare costs an inexorable dollar sum, especially, for reasons unknown or undisclosed, older Hispanic women in the United States. What we can say with some conviction is that this economizing logic is in keeping with an increasingly metrics-based approach to health and healthcare that easily slides into classed, sexed, ableist, and racialized distinctions about how much lives are worth and how much they ostensibly cost to sustain.[53]

Prospecting for Protein Deficiency

Although the evidence for how to treat muscle loss most effectively is decidedly mixed, the vast majority of research, regardless of its findings, concludes with recommendations that veil this ambiguity. On redressing

sarcopenia's costly onset, Goates and colleagues offer typical recommendations: "Based on current clinical evidence, preventive measures such as increasing physical activity and improving nutrition quality (increased protein
intake, use of specialized nutrients that enhance muscle metabolism) should
be considered to help improve muscle function in older adults towards reducing the burden of sarcopenia."[54] In a 2004 study titled "Healthcare Costs
of Sarcopenia in the United States," which argues that "the excess healthcare
expenditures were $860 for every sarcopenic man and $933 for every sarcopenic woman," and that "a 10% reduction in sarcopenia prevalence would
result in savings of $1.1 billion (dollars adjusted to 2000 rate) per year in
U.S. healthcare costs," protein was again the foremost proposed remedy to
offset this cost:

> An initial treatment that may reduce the normal progression of sarco-
> penia in older persons is to ensure that they are eating enough protein.
> Approximately 35% of the older population eats less than the current
> recommended dietary intake (RDI) for protein (0.8 g of protein _ kg–1
> day–1), and about 15% eat less than 75% of this amount. One study re-
> ported that eating about half of the RDI for protein over a 9-week pe-
> riod led to significant reductions in lean body mass in elderly women,
> whereas elderly women who consumed the RDI for protein maintained
> lean body mass. However, it is not known whether modest reductions
> in dietary protein (e.g., 10–15% below RDI) contribute to sarcopenia or
> whether increasing protein intake levels to 100% of the RDI would re-
> sult in an increase in muscle mass in sarcopenic individuals.[55]

This is protein as a preventive measure *and* as an object of speculation. To see
how, it is worth briefly situating these two studies in the political economy of
scientific knowledge production. The 2019 article by Goates and colleagues
was published in the *Journal of Frailty and Aging*, founded in 2012 as an initiative of the International Conference of Frailty and Sarcopenia Research,
which has convened annually since 2011. "Gold partners" of this conference
since 2018 are Nestlé Health Science, a division of a company with a nearly
century-old interest in protein deficiency rooted in the Great Protein Fiasco
discussed in chapter 2. Another partner is Abbott Nutrition, manufacturer
of the popular supplement Ensure. Lead author Scott Goates is listed as a
director at Abbott. The 2004 study on the healthcare costs of sarcopenia in
the United States was funded by the US Department of Agriculture, whose interest in protein we discussed in chapter 3. William Evans, who authored the
widely cited article "What Is Sarcopenia?" and who claims to have been "the

first to describe the condition," has worked across academia and industry for decades, including as Head of the Muscle Metabolism Discovery Performance Unit at GlaxoSmithKline, where he worked to develop new medicines to treat muscle wasting, frailty, and sarcopenia.[56]

These scientific-corporate partnerships demonstrate the reach of neoliberalism into the production of scientific knowledge about the aging population and the individual bodies that it aggregates. Put more pointedly, understanding what is happening in these relationships and the knowledge produced through them is aided by an appreciation of the encroachment of neoliberal logic into the fundamental activities of universities. Comprehending this in full requires a deep dive into the rise of audit cultures, quantitatively oriented managerial techniques, and other metrics of evaluation through which academics are compelled to constitute themselves as knowledge entrepreneurs in ever more corporatized university structures where aligning scholarly and business interests is not only normalized but part of the demonstration of good entrepreneurial subjecthood. It is the political-economic constellation in which scholarly journals and conferences and the interests of the bioeconomy can become coterminous, frictionless entities. It is a contiguous growth that Melinda Cooper and Sunder Rajan, among others, have shown to link the late twentieth-century growth of the life sciences and the ascension of neoliberalism, each developing shared language and conceptual tools around value, reproduction, revolution, and growth itself. And it is an illuminating context in which to situate Irwin Rosenberg's coining of *sarcopenia* in 1988, to see how prevailing scientific and economic conditions were already flowing in an amenable direction of travel.

Although protein supplement and pharmaceutical manufacturers did not invent sarcopenia, the point to emphasize here is that the emergence of this new category of disease cannot be separated from its commercial potential and the voracious market in proteinous commodities that it has come to support. In a 2016 editorial titled "Welcome to the ICD-10 Code for Sarcopenia," proponents of the medicalization of muscle loss in aging, made this link clear. The inclusion of sarcopenia in the WHO's International Statistical Classification of Diseases and Related Health Problems (ICD), and the recognition of sarcopenia as a "disease state," they noted, will "lead to an accelerated interest in physicians making the diagnosis of sarcopenia and for pharmaceutical companies to accelerate the interest in developing drugs to treat sarcopenia."[57] The pressure to obtain research funding (whether from commercial sources such as pharmaceutical companies or from governments and foundations) also plays a role. As one leading researcher on muscle and aging explained in an interview with us when describing why they had transitioned to a focus on sarcopenia, "Quite frankly there's more money, in terms of grants and everything, because aging is a pretty hot area."

Popular supplements targeted at older adults include Swanson's Senior Muscle Retention Protein Powder, a whey compound, and Life Extension's Muscle Strength and Restore Formula, which is high in leucine, a branch chain amino acid that is hailed as vital for muscle building and repair, a paucity of evidence for this claim notwithstanding.[58] Carefully distancing protein supplements from medicines, *Nutrition Insight* claims that the aging consumer demographic is "more resistant to the appeal of a pharmaceutical for every ache and pain," but finds high-protein foodstuffs appealing when they "realize that their muscles don't recover as fast as they once did from everyday active hobbies such as gardening or golfing."[59] Well-established meal replacement brands such as Nestlé's Boost ("When diet's not enough") and Abbott's Ensure ("When food is not enough") that are prescribed and marketed to seniors have also transitioned toward an emphasis on their protein content. In 2018, citing "new research" showing that "more than one in three adults ages 51 and older are not meeting minimum daily protein requirements," Nestlé increased the protein content of Boost from 15 to 20 grams per serving, an increase of 33 percent.[60] In one recent advertising campaign, Abbott used images of active seniors to sell Ensure fortified with NutriVigor, a "unique," proteinous "blend of ingredients" that helps "rebuild muscle" following injury or illness to help people "get back in the game."

Four of six of Abbott US's Ensure shakes are now marketed with protein in their brand names, and the remaining two feature their protein content prominently on the carton. Once used primarily in hospital settings for patients suffering from malnourishment (where they still hold a monopoly in many places across the globe, including the United States and Canada), these drinks are now marketed to healthy people with the promise of a convenient and doctor-recommended path to strength, vitality, longevity, and improved quality of life.[61] Here a focus on recovery with protein as the prescribed remedy, formerly the preserve of high-performance athletes and serious fitness buffs, is extended into more quotidian and moderate forms of movement.

While some contemporary seniors may be wary of medical interventions, efforts to develop drugs to treat muscle loss in aging continue apace. Another partner of the International Conference of Frailty and Sarcopenia Research, the pharmaceutical company Biophytis (tagline "Live Healthier

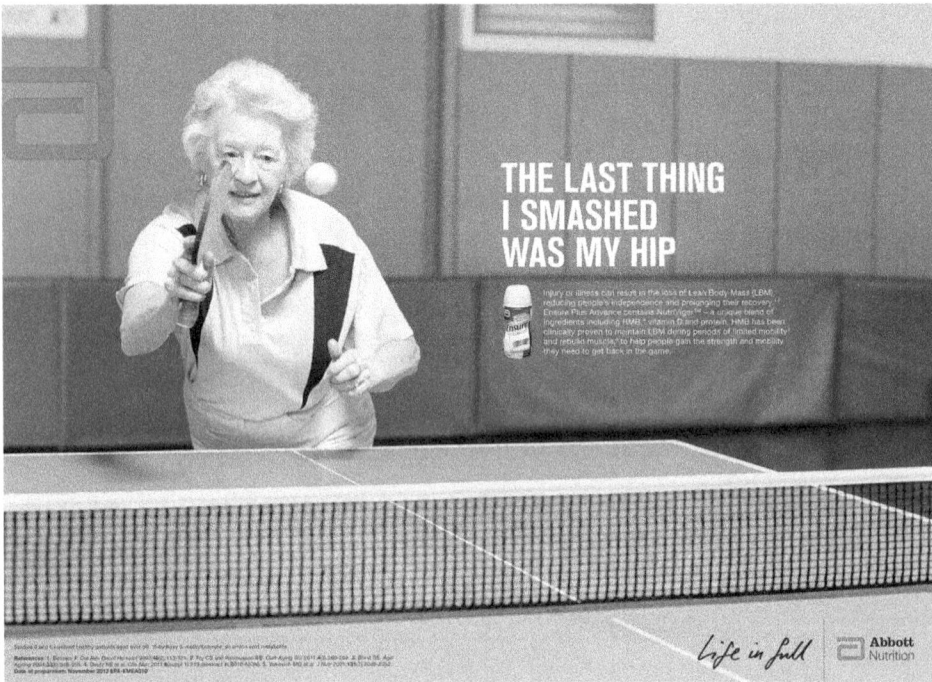

FIGURE 4.1. An advertisement from Ensure's "Life in Full" campaign (ca. 2013), which markets supplements to seniors.

Longer"), is developing a drug called Sarconeos that they describe as "an orally administered small molecule in development for the treatment of neuromuscular diseases. Based on results from cellular and animal studies, we believe Sarconeos (BIO101) stimulates biological resilience through activation of the MAS receptor and may have the potential to improve muscle function and preserve strength, mobility and respiratory capacity in various age-related and muscular wasting conditions."[62] The development of a drug designed to increase protein synthesis, preserve muscle mass, and increase muscular strength and mobility would represent a climax of sarcopenia research should it come to market, and would surely have applications beyond people deemed sarcopenic. At this developmental stage, we can but add it to the web of corporate food, university, and pharmaceutical interests that is increasingly difficult to disentangle, in which the search for problems to which protein is the vaunted solution finds yet another expression in demographic aging.

Conclusion

The homology of protein with muscular growth and overall well-being was formed in the earliest biochemical forays into protein's properties and can be seen in the foundational efforts to establish its ontological distinctiveness and market potential described in chapter 1. The synergy between protein and musculature has since become embedded in cultural norms about what constitutes, quite literally, a healthy body, paving the way for proteinous interventions in bodies deemed to be in age-related decline. In the context of prevailing associations between muscles and youthful vigor, an inverse focus on muscle loss in older adults is a logical one. Yet it was only under particular economic, political, and cultural conditions that the notion of successful, healthy aging through "lifestyle" interventions and other ritualistic habits of the body and of consumption, would become thinkable, let alone open to costing and intervention.

Sarcopenia represents a new category of embodied experience for people to inhabit, one that emphasizes deterioration, and a medicalized category and business opportunity for researchers and prospectors to pursue. While it does not build on a definitional consensus and its etiology remains the subject of foundational debate within the research community, it seeks resonance in the problematizing of aging itself, and demographic aging in particular, as matters of substantive fiscal and corporeal concern. In this light, sarcopenia can be understood as the pursuit of knowledge about aging and muscle decline that economizes human life by conceptualizing the body as a depreciating object. It trades on fears of growing old in late capitalist societies where the social infrastructures of care have been compromised by public spending cuts, where COVID-19 has decimated elderly populations in institutionalized settings especially, and where a decline in muscle mass and function comes at a financial cost to both the individual and the state. It establishes a nebulous threshold for when healthy populations live too long and are apprehended as too costly, for when individual bodies lose too much muscle, and when the extension of life goes too far. At the same time, it produces interventions ranging from physical activity to drugs aimed at protein synthesis and is a vehicle for investment in research and development on protein. The threshold at which a person is deemed costly is linked to the nebulously defined loss of muscle, which together represent the twinned medicalization and economization of muscle loss and their tethering to the process of biological aging. The medicalization of aging and muscle loss pave the way for their economization, which is to say the production of financialized knowledge about their costs and products to mitigate its onset.

Figured this way, it is easy to see why the language and conceptual apparatus for understanding muscle deterioration slips so easily into the language of financialization—and why the age–muscle loss correlation at the heart of sarcopenia is effectively framed as a cost-benefit analysis, where muscular and economic "growth" are conceived and pursued as interrelated categories. Protein promises muscular growth or, at the very least, mitigation against muscle loss. In the former case, where this is about enlarging or sculpting musculature, the conversation stays mostly in the domain of culture, of self-optimization, and the hard yards of exercise. In the latter, it is joined up with matters of government healthcare spending for which individuals are increasingly responsibilized, a context in which its conceptual framework of growth and loss becomes more overtly financialized. Again we see how protein's slipperiness as a concept has made it amenable to multifarious interests and claims; in this instance, protein in the form of pharmaceutical or supplement, depending on the context, is articulated as a commercial solution to the experience of old age. Just as protein and its carnal manifestation as muscle have long been deemed vital and valuable, its absence now entails physical and social worth and potential for intervention and investment.

There is another, final line of analysis to contemplate here that casts sarcopenia research as an especially narrow means of conceiving the issues raised by demographic aging. The equations and logics that inform knowledge about sarcopenia are premised mostly on the notion of an aging population as if this were a singular development. Yet the length of human lifespans fluctuates and is becoming less stable and predictable in the midst of climate breakdown. In *Nomad Century: How to Survive the Climate Upheaval*, Gaia Vince writes that "we are undergoing a species-wide planetary upheaval and it occurs not only at a time of unprecedented climate change but also of human demographic change."[63] Vince notes the consequences for housing and caring for older adults, both in the tropical areas forecast to be worst affected by climate change and also in the parts of the Global North where older adults will soon be supported, or not, by undersized workforces.[64] The matter of muscle mass and its decline with aging is happening amid ecological transformations in which the kinds of care and shelter required to sustain a healthy life will be indelibly affected. Vince's pragmatic vision predicts mass migrations and the emergence of dense cities in places that remain, including those that are becoming habitable for humans on a heating planet, such as the Arctic Circle. In any case, the question of how to feed and house this aging population through the droughts, fires, heatwaves, floods, and pandemics that have already disrupted food production in many parts of the world looms over any discussion about the importance of protein and physical activity in offsetting muscle decline. Physical activity as the default recommendation for all manner of ailments is also compromised on a planet forecast to heat by anywhere between 2 and 10 degrees by the end of the century—a stark prospect that highlights the importance, once again, of seeing matters of health and the body as socially and ecologically embedded. The conditions for performing the athletic labor required to metabolize protein and reap its benefits dissipate with climate upheaval; or, to use Jason Moore's phrasing, the athletic labor of exercise will no longer be "cheap" to perform or prescribe.[65] Questions about how to feed a growing population hungry for proteinous foods are engaged in our epilogue, in which we broach the boom in "alternative" and "green" protein. Before this, we turn our attention to what we call the "muscular manosphere"—the protein market that typically springs to mind when people learn about our work.

5

PROTEIN IN THE
MUSCULAR MANOSPHERE

Supplementation, Self-Optimization, and
Microfascism in Men's Fitness Culture

The self-care advice given to most straight men—inside and out of the Manosphere—goes a little something like this: *"Eat protein! Lift heavy things! Get ripped! Overexert! Get laid!"*
—Maddie McClousky, "Why the MRA 'Manosphere' Isn't Actually Helping Men Cope with Rejection" (2016)

The protein supplement industry as we know it today, with its dizzying array of market segments and global worth of US$21 billion, marks a new chapter in protein's biography, one that is unprecedented in its cultural plurality and economic scope. Since the mid-2000s, sales of sports nutrition bars, bites, drinks, gels, gummies, and powders have skyrocketed, as have sales of every-day foods such as breads and breakfast cereals advertised with the allure of added protein.[1] The growing variety of forms in which a person can consume their amino acids is exceeded only by the growing variety of ingredients and origin stories incorporated into branding repertoires. As of 2024, Walmart, the world's largest retailer, sold 1,169 distinctive products made by twenty brands through its US online store alone.

With food and pharma conglomerates such as Abbott, Hormel, Coca-Cola, Pepsi, and Nestlé monopolizing the lion's share of the sports nutrition market, choice is more constrained than such inventories suggest; choice seems genuine, however, when staged by companies that excel at churning out products with fractionally different biochemical compositions or packaging while making big promises about the unique benefits each provides. For consumers committed to animal proteins, the selection includes whey, casein, egg, beef, and collagen, while popular vegan versions include soy, rice, hemp, pea, quinoa, sprouted grain, and creatine.[2] For those concerned about provenance and allergens, organic, artisanal, grass-fed, fermented, fair trade, GMO-free, hormone-free, and gluten-free options are now ubiquitous.

The booming high-protein economy is fueled by a diverse army of bodybuilders, powerlifters, professional athletes, lifestyle gurus, and fitspiration influencers peddling products to the exercising masses and their aspirants, largely through podcasts and social networking sites such as Instagram, TikTok, and YouTube, but also via traditional media. These sales representatives are pushing not just protein but also a range of cultural and political ideals to which this nutritional category is ever amenable: There are the "strong and sexy" postfeminist empowerment discourses sold through minimalist-design pastel bottles of ready-to-drink protein offered in appropriately feminine (and infantile) flavors like "ice cream" and "birthday cake." There are the solutions to industrial animal agriculture and the climate crisis proffered through vacuum-packed plant-based meats with a protein content that promises to rival the "real thing." And there are the self-disciplinary and self-optimizing responses to the challenges of being a modern man that form our focus in what follows—the regular purchase of oversized plastic containers of protein powder integral to the absolute focus on self-interest and self-sovereignty these codes of manhood require. The primary emphasis on this corner of the market provides a reminder that even as the designer supplements hitherto associated "only with the hardcore factions of male bodybuilders" have become a mainstay of lifestyle routines among a much broader base, men seeking to forge bigger muscles remain a major market and enduring emblem of protein consumption, accounting for approximately 70 percent of all sales according to most estimates.[3]

This chapter pursues amino acid supplementation into perhaps the most obvious site for analysis in a book about protein: men's fitness culture. The modern gym is a broad church, and the gospel of protein widely spread, but there is no getting around the fact that guys of a certain age and body composition are among its most ardent of followers and prominent of cultural signifiers—their aspirations to lift more weight and build more bulk buoyed by a network of influencers with products to sell. In this environment, not any old protein supplement will do. The successful fashioning of a muscular body and masculine identity pivots on the selection of the right brands peddled by the most inspiring expert endorsers, touting not just the most persuasive scientific formulas but entire codes for living. Like other technologies of the self, protein's effect is not simply to facilitate the strongest, leanest, or "healthiest" *generic* masculine body but to circulate particular ideals, promises, and fears about *particular* men's bodies and their place in contemporary cultural politics.

Here our focus is what we call the "muscular manosphere"—a US-rooted, globally resonant, networked men's body culture where life optimization and self-ownership are fueled by protein powder and other lifestyle fixes for masculinities perceived as under threat. We track protein's commodified form as it travels through the lifestyle prescriptions and downline economy of some of the manosphere's most prominent entrepreneurs, including Aubrey Marcus, Joe Rogan, and Alex Jones. The codes for living to which protein is attached involve strong doses of self-responsibility and self-care conveyed in a language that combines—and somehow holds together—scientific certainty, spiritual guidance, gentle reassurance, and brute motivation.

Our discussion highlights the chains of influence through which optimizing creeds and their corresponding supplements are promoted as responses to the perceived challenges of contemporary manhood, where the entreaty to first and foremost look after oneself is a foundational tenet. That said, we depart from established frameworks in which the investment in the fit body is understood as a quintessential manifestation of the self-responsible, productive, individuated, neoliberal subject discussed in the previous chapter. Ongoing evidence for these connections is abundant, but we propose that the muscular manosphere holds a more complex relationship with neoliberalism, one that can emerge only as neoliberalism flounders. That is to say, the emergence of the muscular manosphere is conditional on a neoliberal order in crisis, one grasping to secure conventional modes of exploitation—and control—in the face of growing distrust of market forces, climate and labor precarity, and existential insecurity.[4] In this context, protein-fueled training of the body becomes a salve for a wounded Trumpian masculinity of sorts with links to a latent and emergent "microfascism" in which misogyny and whiteness are inherent.[5] Given the complex and often nebulous politics that circulate through the muscular manosphere, the "latent" and "emergent" are crucial here—the diffuse, affective, relational, and fragile ties that constitute the network representing both its danger and its vulnerability. These are the protein-fueled social and economic relationships that we seek to make visible in what follows.

Masculinity, the Manosphere, and Microfascism

"F*ck exercise," says Aubrey Marcus of "Total Human Optimization" fame. "Exercise is for puppies and babies. We need to train."[6] Only occasionally pictured fully dressed, and frequently styled in caveman chic (figure 5.1), Marcus is a podcaster, self-declared "modern philosopher," and "unconventional fitness junkie" who, with media megastar Joe Rogan, cofounded Onnit, a supplement and wellness brand acquired by Unilever in 2022 for over US$100 million.[7] With this résumé, Marcus is a truly aspirational figure in the muscular manosphere.

For Marcus, Rogan, and their ilk, getting big and strong through hardcore diet and exercise is understood as a worthy end in itself, but it is also mobilized as a way to respond to a broad range of perceived challenges to contemporary manhood. How these challenges are defined varies: At moments, feminism, the denigration of conventional masculine traits and roles, or the profusion of meaningless, sedentary jobs are among the villains to be addressed. At other moments, the sources of men's problems are framed as internal and include spiritual alienation, inadequate self-care, and avoidance of discomfort. In this environment, the successful fashioning of an appropriately masculine self, one that is up to the task of overcoming these challenges, pivots on the selection of the right brands promoted by the most inspiring expert endorsers selling not just the most persuasive scientific formulas but entire codes for living. For Marcus, that code is "total human optimization," an ideal of self-actualization and authenticity that he expands on in a variety of spaces, including: a *New York Times* bestseller, *Own the Day, Own Your Life: Optimized Practices for Waking, Working, Learning, Eating, Training, Playing, Sleeping and Sex*; "Go for Your Win" self-improvement courses, which promise not "easy mastery" but "the truth of a real fight"; a podcast and newsletter; and countless appearances on his fellow gurus' platforms.[8]

Optimization is a capacious category, and Marcus makes the most of this, with teachings ranging from banal advice to eat less sugar and run a humidifier at night to elaborate musings on the "Masculine Archetype" that must be rescued from the accusation of "toxic masculinity" that "misrepresents the totality and

Onnit Founder Aubrey Marcus Is on a (Trippy) Quest to Be a Better Man

From sex fasts and darkness retreats to ecstatic dancing and ayahuasca, Marcus is searching for meaning in the ruins of traditional masculinity.

BY ANNA PEELE   PUBLISHED: FEB 20, 2020

Marcus has a stone garden at his home. Stacking rocks marries Jenga skills, like patience and balance, with brute-strength training.
ANDREW HETHERINGTON

FIGURE 5.1. Screenshot of a *Men's Health* profile of Aubrey Marcus (2020).

potential of masculine energy."[9] For goals both prosaic and lofty, Onnit supplements provide the molecular infrastructure to support their realization. Marcus attributes 90 percent of the company's early growth to podcast marketing—which in practice entailed providing other "brolosophers" with free Onnit products, the promotion of which they could then incorporate into whichever version of ideal modern manhood they are appealing to on their shows.[10]

The emergence of the muscular manosphere is not without precedent, and there are some strong continuities between earlier men's movements and the digital network of today. Recognizing this history, our analysis builds on a longer tradition of research on masculinity and fitness, including ethnographer Michael Atkinson's 2007 exploration of the incorporation of supplementation into the everyday health and lifestyle routines of a group of white, middle-class Canadian men who saw their efforts to build bigger muscles and trim excess fat as a way to "regain, literally, a physical presence of distinction in the workplace."[11] In keeping with a broader literature on the confluence of individualism and fitness culture, Atkinson concluded that these men "construct the consumption of sports supplements as part of a neoliberal, do-it-yourself method of "getting fit" or "healthy."[12] He also argued that their efforts to reshape their bodies through weightlifting and supplementation represented a way to manage their shifting role and status in the face of perceived gains by women and, we would add, the "feminizing" effects of white-collar, desk-based, sedentary labor on their embodiment. Readers will likely recognize the echoes of earlier men's movements in these discourses, including the mythopoetic movement that came to prominence in the 1980s and 1990s under the leadership of poet Robert Bly.[13] Blaming modernization and industrialization for the "feminization" of men, mythopoets advocated for spaces where men could retreat from women and engage in segregated homosocial rituals that would allow them to tap back into their "true manhood" and naturally endowed capacities for leadership. For the men in Atkinson's study, supplements such as protein powder were deemed essential to recovery from physical and social senescence, and their uptake was fraught with gender and class politics. While Atkinson's participants felt threatened by the feminization of working life to their social status, they also sought to "neutralize" this possibility. They did so by downplaying their personal investment in body projects, with comments such as "Who cares? I don't care what people think," which served to distance them from the kinds of body obsessiveness associated with the upkeep of femininity, even as their practices suggested otherwise.

Fitness culture, supplement markets, and the cultural politics of gender have shifted in the nearly two decades since Atkinson's study, albeit in ways that have amplified some of his key findings. Without question, the quest for bigger muscles, undertaken in the space of the gym, remains a core motif and driver of the supplement industry. The gym is where the size or cut of a muscle, the strength of a lift, or the power of a jump are understood to reflect a necessary combination of physical and dietary discipline, where an efficacious protein regimen is embodied and admired through the performance of ideal body shape, size, and capacity, where the ritualized sharing of intricate knowledge about dietary protein—or "protein talk"—acts as a mode of connection and intimacy among clients, and where both word of mouth and direct marketing of protein-rich foods underpins the profit margins of the billion-dollar supplement industry. The gym is also the place where protein is activated metabolically. It is a laboratory of sorts designed to facilitate physical exertion and bring about anatomical and biochemical changes in the body while also recycling the surpluses of agricultural overproduction (as discussed in chapter 3). It is a space where vigorous exercise helps ensure that protein supplements have their desired effect, generating muscle rather than fat, repairing the body's tissues, and allowing further growth to occur. And it is where the biological work that protein is understood to perform is made most visible, thus affirming its association with strength and vitality.

What constitutes a gym has altered significantly since 2007, however, and the same is true for the office—the other social space crucial to Atkinson's analysis. Casualization, digital technology, and a global pandemic have made remote and hybrid workouts—and work—ubiquitous, and the explosion of fitspiration social media has multiplied and dispersed the flow of training norms and ideals far beyond the conventional brick-and-mortar gym and its media equivalents. In today's highly competitive industry, exercise options have burgeoned, with choices ranging from one-size-fits-all big-box franchises to boutique personal fitness services, from bodybuilding and power-lifting iron houses to dojos and boxing gyms, from CrossFit and functional fitness boxes to primal movement and high-intensity interval training (HIIT) studios, and from home gyms, backyards, and streaming services to outdoor bootcamps and obstacle courses. Across these spaces and activities, training options abound, with dumbbells, kettlebells, stones, tires, ropes, weight machines, punching bags, TRXS, stability balls, medicine balls, and plain old body weight among the profusion of technologies in use. Much of this culture is lived online, diffused through Instagram, YouTube, and TikTok, where likes, clicks, follows, and shares ensure fast, wide and incessant circulation of health and fitness edicts.

The interest in supplementation that Atkinson identified among a "select group" of Canadian men has also mushroomed among a much more diverse consumer base and has been affixed to a wider variety of gender and other social norms, even if most converge around some version of self-optimization, and hypermasculinity remains a primary motif. Indeed, the expansion and digitalization of the supplement market might be understood to have intensified and consolidated its hypermasculine core in a way that corresponds with the assertions of a "male backlash" to a shifting gender order.[14] And, crucially, this has happened at the same time as the "flexible" working conditions of late capitalism have become more diffuse, decentralized, and precarious, such that what used to be called the workplace has transcended spatial limits for the postindustrial or knowledge economy worker whose horizons for productivity are ostensibly limited only by digital connectivity or mobile hardware.

It is within this context that the "muscular manosphere," a fitness- and wellness-centric branch of the far-reaching network of online men's communities to which the term *manosphere* has been conventionally applied, comes to make sense as a lens through which to build our analysis.[15] The "manosphere" has been used as a descriptor by both its advocates and detractors to designate an ecosystem of influencers and their acolytes bonded by a commitment to antifeminist, misogynist, right-wing, often libertarian politics, and the idea that "socially, economically and sexually, men are at the whims of women's (and feminists') power and desires."[16] In its contemporary iteration, the manosphere includes long-standing strains of men's and father's rights activism as well as newer assemblages such as pickup artists, involuntary celibates or incels, and the separatist community Men Going Their Own Way (MGTOW). Scholars who write critically about manospheric matters have used a variety of terms—"popular misogyny," "networked misogyny," "gendered cyberhate," "white fratriarchy," "Big Man sovereignty," "microfascism," "viral masculinity," and "bro culture"—to theorize their subjects.[17] Communication studies scholar Karen Ashcraft captures the contemporary manosphere's range and expansion: "With the rise of participatory platforms over the last decade, the manosphere has exploded into a vast and thriving global economy peddling manly grievance in all its varieties: misogyny, homophobia, racism, anti-Semitism, white supremacy and more. It serves up one basic sentiment—*rightful virility, wrongly denied*—in heady brews that range from 'decaf' anti-feminist irony to full-strength virulent violence."[18] While these political currents vibrate forcefully through the manosphere's muscular rendition, we argue that the attraction and malleability of protein as the route to a bigger, more muscular body is such that it links together a broader, messier tangle of shifting orientations and actors—including men who would understand themselves as liberal, or even left wing, and perhaps occasionally even feminist. In other words, not everyone is captured by the manosphere's ideological overtures or affected by it in the same way. But most, if not all, will encounter and negotiate this fulcrum of contemporary networked fitness culture.

Media studies scholar Sarah Banet-Weiser's description of popular misogyny as a "deeply embedded networked context" that refuses to "sit still" is useful for capturing the nature of the relationships that link the diversity of bedfellows that comprise our focus.[19] Like popular misogyny, the muscular manosphere is not coherent as a social movement, and its spaces, discourses, practices, and figureheads tend to be, in J. K. Gibson-Graham's words, "related analogically rather than organizationally and connected through webs of signification."[20] The web that holds the muscular manosphere together is a vision of ideal manhood realized not just by training the body through demanding physical activity and routinized "self-care," and not just by providing that body with healthy food, adequate hydration, and quality sleep, but also by supplementing physical fitness and diet regimens with carefully selected, "scientifically proven" products designed to optimize and enhance performance in all aspects of daily life. In the manosphere, where the desire for self-re-creation and control are paramount, protein offers a way— biochemically and psychically—to internalize and metabolize those desires.

It is here that the fascistic resonances of the muscular manosphere start to emerge. Debate rages as to whether or to what extent classical fascism is discernible in political and cultural formations of the 2020s. Notwithstanding their differences, analyses of fascism's contemporary manifestations tend to converge around the recognition that reactionary, authoritarian, right-wing movements and cultures are on the rise. Among those attempting to theorize this conjuncture is cultural studies scholar Jack Bratich, whose focus is "microfascism."[21] Drawing on the work of philosophers Gilles Deleuze and Félix Guattari and feminist scholars Verónica Gago, Marta Malo, and Lucía Cavallero, Bratich argues that neoliberalism in decline "unleashes and activates microfascism."[22] The *micro-* prefix signals Bratich's interest less in institutionalized and spectacular forms of fascism synonymous with mid-twentieth-century authoritarian regimes than in the often-imperceptible manifestations of eliminationism, nationalism, and violence, though these are blurred and mutually reinforcing in the present moment. In other words, Bratich is focused on fascism as a "fundamentally cultural phenomenon" where aesthetics, industry, embodiment, and sacralization—all processes central to the muscular manosphere—shape the self-understanding and appeal that fascism provides.[23] Bratich sets out to highlight the patriarchal element that he views as core to contemporary microfascism, and to this end his analysis hinges on two key features of fascism drawn from the work of historian and political theorist Roger Griffin: restoration/rebirth and eliminationism. In Bratich's account, the thing to be renewed is an imagined, originary, social order founded in gendered sovereignty, with eliminationism taking the form of a slow diminishment and pacification of women, their capacities, and relevance, especially in the realm of life-giving processes. Importantly for our purposes, Bratich argues that "what is restored is not a mythic past but the performance of founding itself—the capacity to make reality and the authority to establish order"—ideals that come through strongly in the muscular manosphere, where engaging in protein supplementation as a process, and a demonstration of the ability to shape one's material existence, is perhaps as important as the results.[24] In this configuration, protein is a fuel for self-creation, renewal, and optimal capacity. In its appearance as a biochemical substance seemingly divorced from social relations and reproduction, especially as desiccated powder destined for smoothies or shakes, it lends itself to fantasies of patriarchal self-sovereignty, control, and stability.

The diversified, networked, and distributed character of the contemporary gym, diffused across online and offline spaces, is a key condition of possibility for the kinds of loosely coordinated "micro-actions" that move and influence in such a way as to allow "isolated individuals to act together," in this case as consumers of protein and its cultural meanings.[25] Protein flows and connects across such spaces, acting as vibrant currency in the philosophies and entrepreneurial activities that characterize the muscular manosphere, and serving as a point of interaction and mutual recognition between men who might not otherwise interact with one another.

Like other cultures of "aggrieved manhood," the muscular manosphere is compositionally diverse, and the particular masculinities at risk typically remain unspecified in terms of class, race, sex, religion, or ability. Aubrey Marcus is an example of a figure who tends to avoid discussions that are easily identified as political—he is, as journalist Amelia Harnish notes, "proudly anti-political."[26] But what this translates to, as Harnish suggests, is a denial of systemic forces. Marcus made this apparent in a speech he gave to a crowded room at his 2019 Mastermind Weekend, a $795-per-head "transformational" seminar: "Someone said to me recently, 'the problem is all the angry white men.' . . . Angry white men? No, I don't see that. This is one of the big challenges of our time: People who say 'this is a white thing, this is a Black thing, this is a male thing, or a female thing.' All of this is untrue. Don't engage in it."[27] As is typical for this brand of self-help, Marcus's audience was composed almost entirely of large white men who were probably more than happy to hear the message that naming, much less addressing, racism and sexism is best avoided. But the point here is that such conditions do not need to be named. As Bratich suggests, white supremacist heteropatriarchy takes root in desires, relationships, and subjectivities "that don't always crystallize into nameable groups."[28] "Whiteness is not just a demographic or a 'working class left behind' that then turns into a racist bloc," he writes. "Whiteness is an abstraction that galvanizes expansion and control as a matter of course."[29] Whiteness in this formulation is always already gendered, since it is manhood, specifically, that must be resurrected and restored through (bodily) expansion and control.

Thus, while many of the influencers and acolytes who comprise the muscular manosphere may not express or enact misogynist, racist, or other fascist values, they are linked through an orientation to the social sphere and a "subjective structure" defined by the embodiment of "autogenetic sovereignty" or the fantasy of absolute power to self-create and define oneself that is the definition of modern Western manhood. It is this fantasy, Bratich argues, that constitutes "microfascism in the subjective mode."[30] As a vehicle of exchange, and as a "charismatic nutrient" synonymous with renewal, strength, and growth, protein in its supplemental form enables microfascist affects and messages to circulate. In the process, these values become benign everyday notions, often without an explicit or deliberate link to a specific political agenda. In the remainder of the chapter, we track the omnipresent, insidious, and embodied character of these everyday notions, and their attachment to protein marketing through the online interfaces that constitute the manosphere.

What's Your Discipline? Multilevel Marketing and Sovereign Self-Making

"What's Your Discipline?," reads the tab at the top of the online store where Navy Seal veteran, sports nutrition entrepreneur, and global media personality Jocko Willink sells his line of protein powders, amino acid–spiked energy drinks, fish oils, and powdered greens. In bold red and white print set against a black background, six options are offered to consumers: "Combat Athlete," "Industrial Worker," "Active Lifestyle," "Tactically Prepared," "Endurance Games," and "Team Sports."[31] Each category is accompanied by a short explainer. Click "Industrial Worker" if "you're already doing the heavy lifting, rigging, fabricating" and require fuel for "the long haul." Select "Active Lifestyle" if you're "in your chair most of the day, but on the path" and need to "maximize each workout." Choose "Tactically Prepared" if you're "on the trail, tracking your kill" and must "BE READY."

Against a backdrop of machinery and barred windows stands the Industrial Worker. Pictured from built chest down with blue coveralls, heavy-duty gloves, and solid posture, he wears a rusty chain around his neck, his perfectly flexed arm holding a rigging hook in place. Next to him, the representative of Active Lifestyle stands tall and defiant at the top of his deadlift, the bar bending under the strain of eight large plates. A vintage American flag is the only décor in the bare brick warehouse in which he lifts, a world away from the office cubicle where he (regrettably?) spends his days. His thick chest-length beard, the uniform of the January 6 insurrection, is one of many similar sported by the men of Jocko Fuel. Meanwhile, the Tactically Prepared avatar, with camo backpack, khaki pants, and requisite ball cap, moves purposefully across a desert scrub. Click on this tab and you learn how Jocko's supplements are designed with hunters, clay shooters, and target shooters in mind. With names like "Joint Warfare," "Jocko Discipline," "Cold War," and "Jocko Mölk"—the accented *o* a nod, perhaps, to the Norse vibe favored by white nationalist Viking wannabees—the copy emphasizes the role these products can play in building mental focus, a steady aim, and leadership skills.

Despite the highly targeted messaging, all paths through the site ultimately lead to the same set of products, with motivational slogans offering only slightly different versions of the same theme: daily discipline, perpetual improvement, hard work, sustained energy, active recovery. Do hard things. Take responsibility. No excuses, no quitting, only pushing through. Fight ready. From sunup to sundown, build your body, build your brand, build America. The code is summed up in two words, *extreme ownership*, used as the title of Willink's best-selling book on "how U.S. Navy Seals lead and win."[32]

We were first introduced to Jocko Willink by Patrick Wyman, who used one of Willink's podcast episodes as an entry point to his widely circulated Substack essay on bro culture, fitness, and contemporary American masculinity.[33] Wyman argues that while the form of American manhood the Willink brand espouses might seem so exaggerated as to be parodic, it is in reality deadly serious and omnipresent. "If you hang out in combat sports gyms, on military bases, and at construction job sites, or if you watch Joe Rogan clips on YouTube and follow the algorithm down the rabbit hole," Wyman writes, "this type of masculinity—what it means to be a guy—is absolutely ubiquitous."[34] Wyman is quick to point out that bro culture is not monolithic: "Idiot meatheads juiced out of their minds on Trenbolone coexist with doctors of physical therapy who author scholarly studies of movement patterns, absurdly large strongman competitors appear alongside wiry Jiu-Jitsu guys, cleaning-eating vegans and beer enthusiast football fans."[35] Diversity notwithstanding, Wyman argues that these men share in common a commitment to masculinity "rooted in some form of physical capacity," a type of "self-ownership through activity," that is key to grasping how a muscular

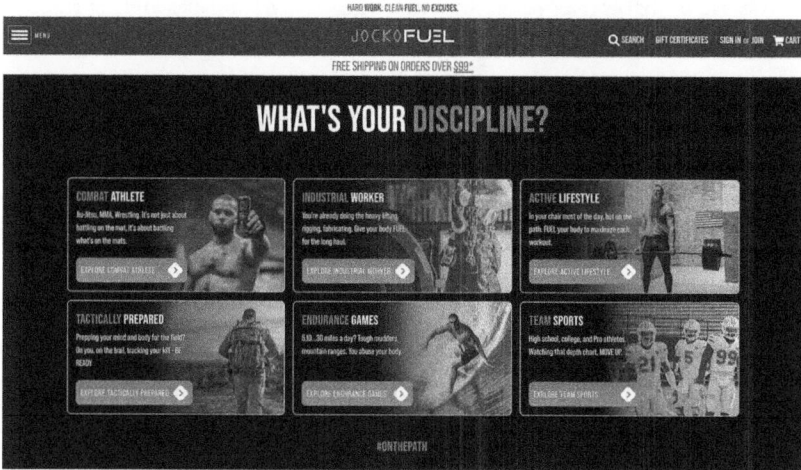

FIGURE 5.2. Jocko Fuel advertisement (ca. 2023) with recommendations tailored to consumer interests and identities.

ethnonationalism has moved from the margins of American political culture to the mainstream.[36]

Wyman situates the rise of bro culture in relation to US empire and decades of perpetual war. He connects the elevation of military service "writ large" during the past thirty years to the direct violence unleashed on foreign, Black and brown others and to the militarization of domestic culture. At the same time that this environment produces entrepreneurs such as Willink, whose battlefield experience is at the heart of his brand, it also generates an eager mass of corresponding consumers—among whom veterans, border agents, cops, and security workers are overrepresented. Wyman posits rising educational gaps, which have emerged between men and women across all racial groups, as a driver for the turn to physical culture, which represents a highly visible way of building status for those locked out of the professional-managerial class. Here Wyman fleshes out the typical bro profile:

[He is] a straight, young-ish (18–40) dude who's kind of into fitness of some kind, whether that's lifting weights, a little jiu-jitsu, or what have you. He probably played sports and currently enjoys watching them. He's familiar with but not super dedicated to video games and likes beer and maybe some weed from time to time. He may or may not have a college degree, but either way has a solid but not extremely high-paying job. He probably lives in the suburbs, exurbs, or a rural area, rather than a dense metro. He's probably but not necessarily white. He's disproportionately likely to have served in the military, and if he hasn't, he knows people—family or friends—who do or did.

Wyman is effective at conveying the vast reach of the online ecosystem "built to cater to Bro Culture," noting that in spite of the diversity of subcultures that comprise it, followers are sorted and funneled into what is ultimately a narrow lifestyle ethos: "It's possible to exist in an entire online ecosystem built to cater to Bro Culture, only occasionally stepping outside its safe and familiar confines, surrounded by videos of jacked dudes talking about squats and MMA and guns."[37] Viewers get sucked in by algorithms that draw them through webs of related material: "If you watch a video on how to improve your bench-press form, you'll be taken to subsequent links on nutrition, powerlifting, bodybuilding, and tactical firearms training, on and on down the rabbit hole of this loose network of social media."[38] Individual figureheads within bro culture are especially crucial in securing these chains of influence: Jocko Willink "might lead you to bowhunter Cameron Hanes, who might take you to the tactical training gurus Pat McNamara or Tony Sentmanat, and from there to former MMA fighter and Green Beret Tim Kennedy," he writes.[39] With tens of thousands of followers—Willink's podcast has 1.6 million subscribers, and each episode is watched on YouTube by an average of 24,200 viewers—these influencers maintain their market share and extend their reach by appearing in each other's videos and podcasts, with "all roads inevitably" leading to Joe Rogan and his "incredibly lucrative empire."[40]

These roads, it turns out, are paved with protein. Not just protein, but a pharmacopeia of powders, pills, teas, oils, drops, and sprays designed to improve gut health, boost immunity, enhance focus, memory, mood, and sleep, and otherwise contribute to the self-empowerment regimes these men promote. Alongside pricey supplements, manosphere influencers sell and endorse a range of New-Age-meets-optimizer-meets-survivalist gadgets, ranging from neti pots and blue light glasses to dehydrated foods and water filtration systems. Even the briefest of sorties into this world are enough to make clear how utterly integral sales—and sales pitches—are to the muscular manosphere. As Matthew Remski, cohost of the podcast *Conspirituality*, wrote in an essay on "bro science," "Everything the men pitch as being good for bodies and minds is a consumer item. Purchasing is the pathway to virtue, and if a bro scientist is pushing an MLM [multilevel marketing] the virtue proposition is increased by networked recruitment. You become a better bro by eating the supplement and you become a super bro by building your supplement downline."[41]

In our journey along these chains of influence, we were taken from Willink and Jocko Fuel to Aubrey Marcus and the vast Onnit marketplace, which includes a "grass-fed" whey isolate "from New Zealand cows," a plant-based protein powder with a much-sought-after "complete amino acid profile," and his flagship "Alpha Brain" capsules, made with the amino acid L-theanine, said to promote a "flow state" and the feeling of being "in the zone."[42] Through Marcus we were introduced to Kyle "Achieve Peak Performance in Everyday Life" Kingsbury, a retired college football player, mixed martial artist, podcaster, former director at Onnit, and promoter of "hunting immersion experiences" targeted at "overworked and painfully domesticated" men.[43] The supplements and snacks Kingsbury promotes include Paleo Valley 100 percent grass-fed beef sticks and Organifi Complete Protein shakes. From Kingsbury we found our way to J. P. Sears, YouTube comedian, wellness satirist turned wellness salesman, and "emotional healing coach" whose endorsements include Ancestral Fuel "animal-based protein powder," cold plunge baths, and Black Rifle Coffee, the veteran-owned brand with the tagline "Fresh Roasted Freedom."[44] From Sears, we were led to Rogan, host of one of the planet's most popular podcasts, optimizer extraordinaire, stand-up comic, mixed martial artist, bow hunter, "psychedelic adventurer," and founder of Onnit. And from Rogan we were transported to Alex Jones, white nationalist, Sandy Hook denialist, and radio personality whose media empire is funded through the sale of his hugely popular Infowars supplements. For Jones, who declared bankruptcy in December 2022 in the face of a $1.5 billion bill for damages he has been required to pay the parents of children killed in the Sandy Hook mass shooting, which Jones claims is a hoax, his supplement business is not simply an add-on to the income he generates through his media empire but the source of its funding, earning him US$165 million between 2015 and 2018 alone.[45] Jones's supplements include Ultimate Bone Broth for "primal human nutrition" and Alpha Power "vital male androgen biosynthesis promoter," among dozens of others.[46]

There are a number of ways to understand the ubiquity and power of protein supplementation in the muscular manosphere, including: the robust economic infrastructure supporting the industry's networked sales and marketing, its homology with muscle and thus with masculinity, and its convenient consumable form and vaunted metabolic speed. All of these features are certainly among the selling points for downline-conscious bros. Less obvious, perhaps, is the reality that for protein supplements to realize their potential requires constant work and renewal through the kinds of physical exertion prescribed and modeled by the manosphere's leading cast. One buys a tub of protein powder in order to metabolize its promises, and doing that requires physical effort over a sustained period of time that is never enough, never complete. The desired physical benefits will duly arrive only if the consumer maintains their regime; any benefits will be lost just as quickly if they don't. Perpetual renewal is needed to replenish muscle, and this lends itself well to more sales and more accumulation of all kinds, including the optimizing discourses of restoration and regeneration that characterize microfascism.

Crucial here is the fantasy of self-creation. As Bratich describes it, this is "a fantasy that nevertheless is implemented in action upon the world to shape reality through order."[47] As protein powder is blended into regimes of optimization, as it is activated as a fuel for self-replication, it becomes a key ingredient in the masculine mythmaking and self-making that gains its desirability because it allows its subject to form through separation from its "material conditions, interdependence, and sociality."[48] Returning to themes discussed in chapter 3, we can see how a supplement such as whey protein appears in a form that is isolated from its circumstances of production and from its worldly relations. Intended only to serve the optimized health of the individual consumer, the socio-ecological relations and economic imperatives that underpin the creation, circulation, and consumption of protein supplements are nowhere to be seen in desiccated powders or self-made men who make up the muscular manosphere. "Total human optimization." "Extreme ownership." According to Bratich, the "flight from materiality" allows abstract values of this sort to emerge; these values then "return as impositions on the world through hierarchical ordering" of sex, race, age, and so on represented and energized through the manosphere.[49]

Readers might question our choice to analyze the manosphere through the lens of microfascism, wondering if this generalizes or exaggerates its political leanings, exclusions, and violence. But Bratich's point is that microfascist cultural politics is "composed before being organized, directed, even named," that it emerges nonlinearly, and that it can have effects "even if they don't 'creep' to another stage."[50] For Bratich, microfascism as the cultivation of subjectivity is ordinary, pervasive, latent, and multiscalar. It is "'in our heads' but also in our bodies. . . . Better yet, it's in the embodiments and affective armor produced within our relationships."[51] This means that patriarchy or autogenetic sovereignty are not individual propensities, or even products of abstract systems such as "patriarchy" or "fascism." Rather, microfascism is the "realm of composition" and thus requires an analytical focus on the "collective and assembling formations."[52] Bratich writes, "The contemporary variations of these patriarchal pacts, specifically in networked warrior bands, will be key to understanding microfascism."[53] Thinking with Bratich in this way allows us to remain alert to the ways that microfascism infiltrates desires, values, and practices that don't always develop into identifiable groups or organizations, instead embedding itself in relationships and subjectivities across the political and cultural spectrum.

Indeed, the leaders of the muscular manosphere exist less on a political continuum and more in a tangled knot, which means that self-proclaimed "bleeding heart liberal" Rogan regularly plays host on his podcast to the far-right's Alex Jones, interspersing the libertarianism and (surface) nonpartisanism that permeate the manosphere with occasional dashes of liberalism and generous servings of Trumpism and the Proud Boys. His endorsement of Trump on the eve of the 2024 election on the *Joe Rogan Experience* podcast, however, was an unambiguous declaration (a declaration that was as much an endorsement of Elon Musk as of Trump, around the time of their fraught bromance to which we return in our epilogue.) For the main, though, Rogan's capacity to occupy an ambiguous political position across multiple belief systems is key to his hugely popular brand of cultural politics.

For a sense of how these belief systems merge or are expressed through mind-body cultures and in the podcasts, YouTube channels, and other networked spaces of the muscular manosphere, think: fitness for survival, combat, and self-defense; target shooting, bow hunting, and mixed martial arts; kettlebell flow, primal movement, and stone lifting; paganism, primitivism, and New Age spirituality; poets, mystics and shamans; mushrooms, marijuana, and psychedelics; organic foods, clean foods, and wild foods; big-ag-, big-pharma-, and mainstream-science-skepticism; "free thinkers," conspiracy theorizers, and contrarians; self-help, self-care, and self-discipline; market fundamentalism, entrepreneurialism, and hyperindividualism; rugged heteronormativity, hyper-ablebodiedness, and eternal youth. Think "Buddha-Bros," "psychedelic masculinity," "tactical culture," "combat culture," "rationalized body culture," "podcast bros," "Bitcoin carnivores," "conspirituality," "hybrid masculinity," and "beta masculinity." Think appropriation and exoticization of Indigenous motifs alongside the erasure of Indigenous peoples and struggles. Think fetishization of nature, wilderness, and the outdoors unmoored from colonial histories. Think demonization of indoor, sedentary living, industrialization, and automatization but void of systemic critique. Think uniformly pretty white wives, partners, and nonexclusive lovers (polyamory is big in the muscular manosphere) who, if they appear as active subjects at all, are fiercely domesticated raisers of these men's children, but also able to shoot a gun, wield an axe, and swing a hefty kettlebell. Think Willink, Marcus, Kingsbury, Sears, Rogan, and Jones, but also Paul Chek, the Liver King, Aaron Rodgers, Andrew Tate, and even Jordan Peterson. Peterson might not be built like a bull, and to our knowledge he doesn't sell supplements, but he is nonetheless known for his all-meat diet and other disciplined habits of living celebrated by more vibrant and dare we say "manly" representatives of the manosphere. His approach to self-help—outlined in his book *Beyond Order: 12 More Rules for Life*, successor to the best-selling *12 Rules for Life: An Antidote to Chaos*—echoes a number of key themes that reverberate through the muscular manosphere, where militaristic commands to "stand up straight with your shoulders back," "set your alarm clock," "drink your water," and "make your bed" are offered in conjunction with more warm and fuzzy advice to "treat yourself like someone you are responsible for helping."[54] These strains of advice do not necessarily exist in tension but instead operate as examples of the "ruthless self-love" (or self-optimization, self-improvement, self-discipline, self-reliance, and self-cultivation) these figureheads promote.[55]

The über-individualism of manospheric culture bears emphasizing. As Remski writes, "In the rugged country of bros, there are no communities, but rather collections of homogenous, self-contained, self-responsible individuals. They mirror each other's strong and silent bro-ness. They owe each other nothing but affiliate links."[56] While the extent to which these men are explicit about their product sales varies, the economy of supplementation that underpins this network often goes unacknowledged, perhaps because discussing dependence on their audiences' financial investments might undermine the individualism that is so crucial to their brands. Likewise, income or employment are infrequently mentioned by manosphere figureheads who tend to ritualistically reassure their followers that money does not equate to a meaningful life, even as their obvious or widely known wealth is part of their popular appeal. Their key take-home message is that the Willinks and Marcuses of the world are not waged or salaried workers with a boss or a manager overseeing their productivity. Usually portrayed as free of any institutional or organizational affiliations, much less obligations, they are their own business. They are entrepreneurs of their own lives, both literally and figuratively, their sphere of action and influence extending beyond anything recognizable as work to "human existence in its entirety."[57] This notional freedom is peddled to workers active in the gig economy and other forms of flexible, precarious labor, where being one's own boss is packaged as the realization of neoliberalism's promise of entrepreneurialism for all. By contrast, dependencies of various kinds are precisely what the flexibiliza-tion of labor is intended to conceal and exploit: In the United States, most of the work categorized as flexible or precarious is undertaken by women (especially those with children) or Black and Latino men, especially in the South, while the goal of flexibilization is one of neoliberalism's tactics for eschewing the full economic cost of this workforce and of reproducing social infrastructure.[58]

The muscular manosphere's endless exhortations to control oneself with the goal of singular flourishing leave little room for relationality of any kind. Only individual subjects are recognized within this paradigm because, be-yond the shared bounty of multilevel marketing, interdependence might require incomprehensible kinds of compromise, including limits on the free-dom to accumulate wealth, shoot guns, eat meat, burn fossil fuels, and so on. This came through clearly in the early months of the COVID-19 pandemic, when supplement purchases skyrocketed, justified and promoted by the "bro science" truth-bonding that provides hours of content for manosphere pod-casters selling invulnerability to their fellow renegades and freethinkers.[59] Aubrey Marcus's prescription, penned prior to the pandemic and endlessly amplified since 2020, is typical: "When you get sick, it is not the virus that gives you the symptoms; it is your immune system."[60] For Jocko Willink, who joins Joe Rogan as a promulgator of the idea that obesity is the taboo and suppressed reason for the United States' high COVID death rates, the advice is similarly simple: "Be on a good diet and get outside and get healthy. It's pretty straightforward."[61] As Remski writes,

> The target of bro science is not public health but sales—of supple-ments, self-empowerment regimes, confidence-building, all proposed as totalizing and perfect answers. This is why bro scientists must re-ject the reality of a novel virus they cannot solve. . . . The banal health products of bro science can never be oriented towards the commons; there's no money in it. Everything they pitch as being good for bodies and minds is a consumer item. . . . For the bro scientist, health is the outcome of personal choices not public policy, which is why he cannot conceive of the logic of masks or vaccines, which are not useful as per-sonal choices but as social commitments.[62]

While some figureheads in the muscular manosphere accept the existence of COVID, they share a commitment to minimizing its risks and overstating the extent to which an individual can prevent or fight off an infection through fitness and lifestyle practices. This is about kindness to oneself, perhaps, but not kindness toward others or a sense of collective vulnerability, solidarity, or relationality. The manosphere's emphasis on "clean eating" and wild foods (consumption of highly processed supplements notwithstanding) makes sense in this context. Much like ruthless self-care and immune-boosting supplements for staving off COVID, clean eating offers a way of removing all externalities—like poor-quality food derived from a broken food system—that might interfere with self-optimization and renewal. Here there is an absolute focus on what benefits you, and anything that interferes with optimization or with your quest for sovereignty must be struck out. Recognizing that one is in relationships with others mean ceding some control and in the manosphere the message is unequivocally that you shouldn't get bogged down in what you can't control; that help from others is unreliable and undesirable. All you have is you and your carefully formulated, scientifically proven stock of supplements. In this context, the interconnected discourses and practices of protein supplementation, COVID minimalism, and clean eating can be understood as manifestations of the struggle for the renewal of sovereignty and as techniques for managing, at an individual level, multiple systems in crisis, from food to public health.

Conclusion

If our argument were to stop here, with individual responses to systemic failures, it would replicate a now classic critique of neoliberalism and the production of the responsibilized subject. The muscular manosphere is undoubtedly caught up in the neoliberal project of self-regulation, self-betterment, and self-optimization, and premised on the precarity of flexible modes of accumulation. But we have also tried to show that participants in the manosphere are buying and selling confidence-building regimes and fuels for self-creation and renewal that profit from the fallout of this emergent conjuncture, where success seems harder to come by than before. This vision of disenfranchised men on the path to renewal, rooted in a sometimes latent gender and race supremacy, is not incompatible with neoliberalism, nor is it entirely new. Rather, it is its networked character and its enmeshment in the supplement industry and the project of building bigger, stronger bodies, often for undefined tactical purposes, that holds our analytic attention. While the insurrectionist spirit of wounded Trumpian masculinity stands as a vibrant market for those who profit from the loss of self-confidence these men seem to experience, we hope to have conveyed the messy tangle that characterizes this networked world, pointing to its microfascist vibes, potentialities, and affects rather than reducing it to the here and now of fascism as state formation or organized, intentional form. The power of protein in this environment lies in its capacity for optimization, renewal, and growth—fuel for a fantasy of corporeal sovereignty that belies its utterly relational essence.

This chapter has responded to a much broader question about the kinds of sociality that cohere around protein in the early twenty-first century. In the muscular manosphere, techniques of somatic self-making are dispersed through digital networks of accumulation, where they become allied to nutritionist understandings of protein and both dominant and emergent modes of identity and desire. The muscular manosphere can therefore be understood in continuity with protein's persistent connection with fortifying men's bodies, a connection heightened under neoliberalism. At the same time, protein in the manosphere is a site for expressing fantasies of autogenesis that are indicative of neoliberalism's purported decline and the concomitant rise of microfascism. With Jack Bratich and Sarah Banet-Weiser's arguments as our guide, we can see how the appeal to neoliberal values increasingly struggles to "secure subjects within its political rationality," a claim they use to explain the rise of "networked misogyny," or the "interconnected nodes in a mediated network of misogynistic discourses and practices."[63] In this sense, protein can be viewed as fortifying the vestiges of neoliberalism's hold on contemporary subjectivity while simultaneously fueling emergent microfascisms.

Of course, amino acid supplementation and immersion in fitness culture do not guarantee entry or indoctrination into the manosphere's political field, but the connection should not be underestimated. A recent survey of sixteen- and seventeen-year-old boys by UK antifascism charity Hope not Hate found that more than eight in ten respondents had encountered content from Andrew Tate, the former champion kickboxer turned misogynist manosphere kingpin.[64] The algorithmic logics of platforms such as TikTok and Instagram need reckoning with to explain this, as does the shifting role of gender in cultural politics and the rise of the alt-right with its various iterations of fascism and misogyny. Protein is less an explanation than a vector in these networks, a substance that flows easily through these conjunctures with the promise of enrichment of all kinds. Protein's allure remains evasive and shrouded in mystique, even as the scientized language of metabolisms and amino acids imbues its nutritional power with ever greater objectivity and authority.

EPILOGUE

Between Meat Protectionism
and Alt-Protein Futurism

The supplement industry has come a long way since Justus von Liebig first patented his Extract of Meat in the late nineteenth century with the conviction that it would provide a necessary boost to the population's health. That dark brown syrup with a "powerful beef aroma" turned out to be lacking in nutritional value, but certain continuities are noteworthy.[1] Liebig's concern with bolstering the fitness of men in industrial and military workforces by concentrating and commodifying protein in supplemental form evinces a biopolitical imperative that persists today, as does the interest his extract inspired in physicians hopeful about its curative properties and potential for restoring the tissues and strength of the sick and infirm. Then, as now, protein was allied to a shifting category of health where governmental rationalities regarding the management of risk and responsibility converge. Then, as now, particular bodies, their labor, potential, and waste, were integral to protein's commodification, even as the ideals animating its production, distribution, consumption, and regeneration have become more diverse and diffuse than they were in Liebig's time.

In this book we have sought to chart these ideals and the political imperatives, bodily desires, cultural anxieties, environmental risks, and opportunities for accumulation to which protein consumption has since been adhered. Crucial to our argument is the proposition that protein's status as a nutritional superstar is neither new nor self-evident. Rather, protein has been produced as an über-nutrient over the course of two centuries, a trajectory we have traced by focusing on moments of heightened interest in its seemingly endless potential to address problems of global importance, from food security to population aging to manhood in decline. We have shown that the things and processes that travel under the name *protein* are organic and life-giving but also socially constituted, forged together into what largely passes for a distinct biochemical category in the face of ongoing epistemological and ontological uncertainty and change. In showing how protein's capacities are put to work, and in following its travels as a valuable commodity, lively ecological entity, abstract nutritional input, and facet of bodily composition, we have situated it in the complex political, economic, and cultural contexts that represent both its conditions of possibility and the sites through which its cultural meanings and material impacts are manifested. Our chapters have been organized as a series of episodes demonstrating protein's adaptable, relational essence, each an attempt to show how this nutritional superstar has been animated over time—by the epistemological contestations of early organic chemistry, by imperial expansionism, by monetized solutions to environmental harm, by the neoliberal accounting of the body's senescence, by the fraught cultural politics of masculinity, and now, as we suggest in what follows, by a meat-first populism that regards protein innovation as an existential threat.

In tracing these histories, we have sought to make protein's metabolic forces and cultural valences visible and tangible, revealing how its arcane and abstract character gets enlisted in the service of all sorts of investments, messages, and desires. In so doing, we have been careful to avoid the suggestion that protein's rise has been seamless or linear, or that its hallowed status is inevitable. Rather, we have argued that even—indeed especially—at moments of peak traction, protein's potential has been contested within and across scientific and social domains. The near-decade during which this book has come to fruition is no exception. While this time span represents a small window in the long history of protein, it has been an especially turbulent one in our subject's biography. We have watched in real time as the meanings and materialities of protein have mutated and shifted, its grip on our nutritional imaginations becoming at once more expansive and more fraught. Our epilogue elaborates on this tension, exploring how and why the protein supplementation market continues to grow even as the science underpinning it remains contested, even as its manifestations old and new are challenged by competing visions of climate and sustainability, and even as answers to the seemingly simple dietary questions "How much?" and "What kind?" become more desired and more difficult.

THE IDEA THAT HIGH-PROTEIN diets aid in weight loss, stimulate muscle growth, and generate energy continues to hold sway and is borne out in market research predicting that sales of protein powders and protein-enriched drinks, snacks, and packaged foods should see a 8.65 percent compound annual growth rate between 2023 and 2028 as the market reaches a value of US$50.2 billion.[2] Sales of plant-based high-protein goods are driving growth, but so are sales of animal derivatives, the expansion across both streams enabled by a movement of consumers away from earlier generations of sport nutrition snacks and beverages that did not include protein. This is to say nothing of the market for whole meat proteins, the global consumption of which is projected to increase by 12 percent between 2023 and 2032.[3] Heightened interest in protein supplementation now extends beyond the food and beverage sector to include personal care and cosmetics, where the market for whey-infused facial treatments is thriving.[4] Clinique is among the major brands that highlight the use of whey in their promotions. This is "not your smoothie's whey protein," their ad copy counsels, reminding consumers that their version of whey builds not muscle but skin, offering a "de-aging triple threat" that promises to plump, smooth, and rehydrate the complexion.

Even as protein commerce thrives, the research on a long list of questions related to amino acid intake and health remains far from settled. The lack of consensus is apparent in work on the impact of protein consumption on weight loss, cardiovascular function, and healthy aging—the kinds of chronic disease and lifestyle issues that have come to form the focus of research in the Global North in the past half century. Long-standing tensions, akin to those that have characterized nutrition science since Liebig began his work in the mid-nineteenth century, also continue to reverberate. A strong constituency of experts now suggests that the official recommended daily allowance (RDA) may underestimate the amount of protein people (especially the elderly and the critically ill) need by as much as 50 percent, claiming that concerns about the adverse health effects of higher protein intake have been overstated. Scientists in this group often point to the weaknesses in the methodology used to determine the RDA and emphasize that the RDA is designed to estimate the minimum amount of protein that must be consumed to prevent nitrogen loss, not the optimal range of protein intake. At the opposite end of the spectrum, experts claim that the average diet of Americans and Canadians contains too much protein, highlighting studies that provide evidence to counter the findings driving the idea that older adults, especially, are not eating adequate protein. The reasons for their concerns include the impact of high-protein diets on renal and bone health, on rates of overweight and obesity, on risk for cancer, and on the environment and climate change.[5] Scholars working on questions related to malnutrition, food insecurity, and famine in the Global South are also divided. One can see in this literature echoes of mid-twentieth-century debates (discussed in chapter 2) about whether any special emphasis should be placed on protein deficiency compared with deficiency in other nutrients, or in the context of overall caloric intake, or in relation to systemic forces such as conflict, climate change, economic precarity, and forcible displacement.[6]

That an overarching consensus remains elusive is unsurprising. The metabolic contingencies of protein, the bodies that ingest and process it, and the cultures and systems in which these bodies live and breathe represent a kinetic dynamism that does not lend itself to epistemological certainty or straightforward dietary advice. The different worldviews and commercial interests that shape research on protein metabolism and nutrition only compound the difficulty of navigating a clear path through the evidence. How this evidence will be weighed in the forthcoming review and update of the protein dietary reference intakes in the United States and Canada remains to be seen.[7] The strength of the divergences that characterize the field will not be easy to bridge, even though the science behind the headlines in the "protein wars" is more nuanced and tentative than truncated soundbites about the consequences of eating too much or too little protein suggest. Indeed, if there is a consensus in the literature, it is that the identification of a generalizable optimal intake, even for application within relatively homogeneous populations, is unlikely.[8] This means we may have to be satisfied with "It depends" as the answer to the apparently simple question "How much protein should we eat?"[9]

This is a question we are inevitably asked to address when we speak about our research. We have been struck by how consistently we are drawn back to it even as our analysis problematizes its very foundation. Our attempts to turn the question around, to dislodge its assumptions, and to emphasize its irresolvability seem not to satisfy those holding out for a prescription and some reassurance that they are doing the right thing as they experiment with their diets and manage their health. The belief that a person can and should identify just the right dose and type of protein for peak well-being reflects the resilience of nutritionism and the hold of scientized principles on what and how we eat. It is also an expression of the molecular biopolitics that compel relatively affluent populations to optimize their health at ever more personalized and intricate scales. And, as we contended in chapter 5, it is further linked up with fantasies of corporeal integrity, autonomy, and sovereignty that are taking increasingly troubling forms.

In offering ways to think about protein beyond the quest for nutritional precision and unimpeachable veracity, we are in dialogue with a body of scholarship that seeks a path out of the polarizing, confusing, and normalizing nature of contemporary dietary discourse. Alissa Overend's argument against "singular food truths," which situates the push and pull of nutritional advice in a broader cultural moment where the power of objectivity and empiricism to shape public perceptions has dramatically declined, is particularly relevant. Overend doesn't deny the damage wrought by a manipulative food industry, or the impact of bad science on people's health, but sees little value in an approach that seeks to separate "food fact from food fiction," as if there is a route to eating well that is uncontaminated by social and political forces.[10] Thus, rather than debunking or clarifying nutritional edicts that encourage consumers to veer from one dietary magic bullet to the next, Overend invites us to imagine what would happen if we were to abandon the endless and futile pursuit of singular truths about healthy eating. While Overend distances herself from what she calls a "Trump-style assault on fact" in a brief aside, she argues that we must take seriously the influence of emotion and belief on popular opinion and reject better reason or more counterfacts as solutions to impasses of a variety of kinds, including "what to eat."[11] In making this case, she joins other scholars advocating against reductionist, perfectionist, and anthropocentric attitudes to food in favor of ecological, qualitative, and messy approaches that, in Lisa Heldke's words, "view foods not as substances, but as *loci of relations*."[12]

SUCH A PERSPECTIVE RESONATES STRONGLY with ours, but we are writing this conclusion at a moment where desires for more singular guidance about protein consumption have come to feel especially weighty, in part because the Trumpian context inflects them in ways that cannot be set aside. Protein science has always been contentious, and it has always been practiced and debated beyond the realm of the laboratory. It has also, from the start, been infused with political ideologies and linked to large-scale social upheaval and transformation. But in recent years, the regular and inevitable uncertainties and contestations that constitute the scientific process have been amplified and diffused across multiple spheres and platforms in ways that demand critical attention.

The cacophony of voices, perspectives, and interests engaged in contemporary "protein talk" epitomizes the decentering of traditional scientific expertise and authority that characterizes a post-truth, post-trust world. That protein knowledge is developed and circulated across diffuse locales does not mean it is democratized. The influence of media powerhouse Joe Rogan, who has built his podcast empire through the development of new knowledge pathways centered on health, makes this abundantly clear. The percolation of Rogan's preferences for high-protein food sources through public consciousness and into conversations about our work has become increasingly evident over the period in which our research has unfolded.

As experts and avenues for information have multiplied, and anxieties about the food system, big pharma, and the power of corporations have intensified across the political spectrum, protein has joined the orbit of other objects of contention in the contemporary culture wars. Like antiracist education and transgender humanity and rights, protein—what it is, what it should be, and who might be stopping you from eating it—has become embroiled in the animosity of the "social industry," where its contentiousness is its currency.[13] For protein, that currency lies in its multiplicitous form, in its capacity to appear or to be invoked as meat, or powder, or dairy, or legumes, or cells, or insects, or muscle, in its ability to stand in for the multiple crises and struggles that mark the contemporary world.

Take the figure of the "soy boy," a prominent folk devil in right-wing conspiracy circles.[14] The "soy boy" epithet is used to describe men who are perceived to be lacking in typically masculine qualities and is linked to long-standing sexist-racist tropes about "effeminate rice eaters" (discussed in chapter 2) and, by contrast, to white supremacist discourses that associate dairy milk consumption with racial purity.[15] By intimation, soy boys (or "cucks" or "low-Ts" or "nu-boys") are liberal or left in their politics and victims of the alleged but unproven feminizing properties of the phytoestrogens found in soybeans as they eschew the meat- and dairy-heavy diets to which real men are supposed to adhere.[16] Bill Gates as a stand-in for environmentally conscious Silicon Valley liberals is another favorite villain. Gates's investments in cellular and insect protein development, and his statement in a 2021 interview that "all rich countries should move to 100% synthetic beef," provide the grounds for the claim that he is part of a global elite that plans to end livestock farming, enslave American people, and force them to survive on bugs.[17]

While substituting meat with substances derived from insects represents a small subset of efforts to address the role of animal agriculture in the climate crisis, the civilizational threat posed by the imposition of bug-based foods has become a mainstay of right-wing populist discourse in the 2020s.[18] These fears were stoked in director Bong Joon Ho's 2013 movie *Snowpiercer*, in which humans survive an apocalyptic frozen world by living on a train in perpetual motion. The train's passengers are stratified by their class status, with the masses residing in the tail and the front reserved for the few. During a rebellion, the advancing masses are incited by their discovery that the protein blocks on which they subsist are derived from ground-up cockroaches bred on human waste. Being fed insects is a crowning indignity, more so as those in the front eat sushi and meats derived from the train's livestock supplies. Protein sufficiency is not the question; rather, it is the type and provenance that act as symbols of class power, such that eating insects is a dehumanizing revelation for the masses, tantamount to the other brutal indignities they suffer on board. The overt class struggle over different forms of protein is notable, as is the fact that protein as a category survives the end of the world.

As fears of meat alternatives are taken up in popular cultural forms and embroiled in the culture wars writ large, love of meat, red meat especially, has become a "general symbol of conservatism," and the defense of it a battle cry against (real and imagined) efforts to address climate change.[19] This is why, in a 2022 election, a Montana Republican candidate for office made a spectacle of branding a calf. It is why, in 2024, both houses in the Florida legislature passed a bill banning the sale of lab-cultivated meat. And it is how *The Eggs Benedict Option*, a 2022 best-selling book authored by the Raw Egg Nationalist, breathed new life into old tropes about the crucial role of animal protein in protecting "Western" values from immigration, feminism, and globalism.[20]

These are tropes with far-reaching ideological reverberations. On the cover of a September 2020 issue of *Texas Monthly*, the war against beef is symbolized by several red darts piercing a cut of raw steak, the projectiles presumably thrown by animal rights activists, health and diet enthusiasts, and environmentalists whose activities threaten the viability of the livestock market.[21] Notwithstanding the more nuanced analysis presented in the article itself, the idea of an embattled and failing cattle industry circulates with

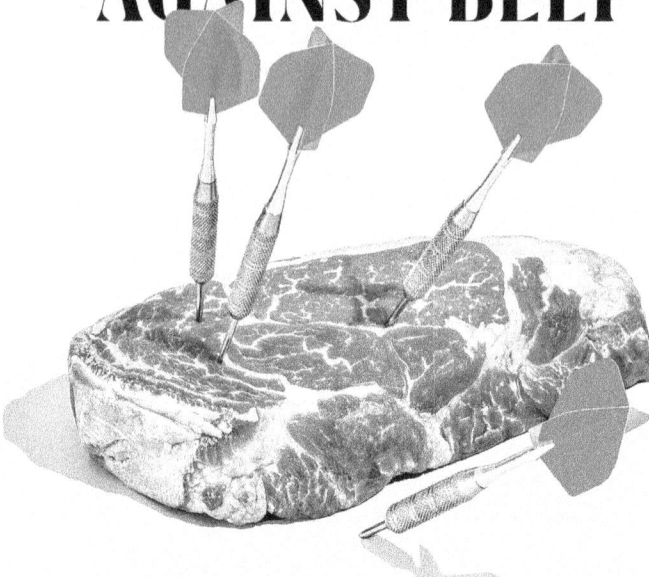

FIGURE E.1. September 2020 cover of a *Texas Monthly* issue focused on innovative cattle ranching practices amid controversies around sustainability in the American beef industry.

the support of an industry that is using tobacco and fossil fuel company tactics to downplay the impact of animal agriculture on the climate, attack the science highlighting this connection, and undermine the ecological efficacy of alternative proteins. These campaigns have had some success. Recent upheaval in the world of alternative protein investment, research, and development, and the moderation of expectations about the commercial viability of industrial plant-based and cellular foods, has a number of antecedents, but the struggles of the new generation of plant-based meat products such as the Beyond Burger have undoubtedly been fueled by a meat-industry-backed campaign that aimed to paint these analogues as packed with fat and hidden additives and therefore less healthy than beef or pork.[22] In their especially incisive analysis of this conjuncture, Jan Dutkiewicz and Gabriel Rosenberg observe that "conservatives' darker fantasies aren't just about threats to a dietary staple but about threats to the liberty, bodily integrity, and masculinity of American men."[23] One particular paragraph is worth quoting at length:

> That's the essence of the right's culture-war metabolism: It transforms the energy of grievance into the substance of reactionary policy. The culture warriors win not when the war is over (it can never end) but when they scuttle any possibility of solving actual problems. They and their political allies frequently seem to want to destroy the capacity of the state to regulate the food supply with fairness, safety, and sustainability in mind. Ironically, it is this very absence of government regulation that produces a nightmarish outcome not far from the conspiracy theorist's vision of mandatory bug bricks: a dystopia in which most food comes from a handful of shady megacorporations that sell chemical-laden flesh of mysterious provenance. Scary? It already exists. It's what they sell in the meat case in every grocery store in America.[24]

The 2024 election of Donald Trump to a second term in office on a platform that includes deregulation of the food and agriculture industries, and an open disregard for efforts to tackle the climate crisis, is unfolding with predictable consequences for health and safety. Trump's appointment of environmental lawyer, antivax conspiracist, and agriculture system critic Robert F. Kennedy Jr. as secretary of Health and Human Services is relevant here. This appointment might seem counter to Trump's antiregulatory agenda, given Kennedy's opposition to ultra-processed foods and other consumables he blames for the American public's poor health. But Kennedy is also making good on his promise to dismantle vital public health research infrastructure and programming. These tensions reflect how Kennedy's foray into federal politics has made unlikely bedfellows of people from across the political spectrum who are concerned about corruption and manipulation in the food and drug systems, pulling new recruits into the orbit of grievance politics that fuel culture war metabolism.[25]

Lest we risk pitching the meat populism of today's political right as the polar opposite to the technoscientific pursuit of alternative proteins in laboratories or vertical farms, it is worth highlighting some of their synergies in the contemporary United States. The recent lunge (real or perceived) of Silicon Valley to the political right—epitomized by tech billionaire Elon Musk's short-lived role in Trump's inner circle—has attracted much attention, especially since Musk's empire has been built around the manufacture of electric cars and, more recently, his purchase and rebranding of Twitter.[26] These ambiguities are also apparent in Musk's history of downplaying the role of the beef industry in climate change at the same time that his company, SpaceX, collaborates on research investigating the viability of growing cultivated meat, albeit beyond Earth's atmosphere. Musk exemplifies the kinds of tensions and paradoxes that characterize this moment, as does the fact that some alternative protein developers have joined efforts with established meat conglomerates such as Cargill, Maple Leaf Foods, and Tyson, that are rebranding themselves as "protein companies."[27] Adding to the puzzle is the shrinking distance between the rhetorical strategies of the legacy companies and outfits such as Beyond Meat, which has trademarked the phrase "The Future of Protein," noting that part of their vision is to "re-imagine the meat section as the Protein Section of the store."[28] In this conjuncture, alt-right, grievance-fueled "muscular capitalism" and defenders of meat, on the one hand, and liberal tech green capitalism and promoters of alternative proteins, on the other, differentially cohere around the climate crisis. While Trumpian populism veers between open dismissiveness and outright denial of environmental breakdown, alternative protein production is premised on its real and pressing consequences. But our bigger point is to observe how protein has become adhered to the climate crisis, even as meat protectionism and alt-protein futurism expound starkly different visions of a planetary future. These ambiguities, synergies, and uncertainties about the future return us to the proposition with which we opened this book: that protein's elusiveness, dynamic, contingent, multiplicitous, and adhesive qualities are key to understanding its power.

PROTEIN'S UNRULY NATURE FACILITATES ITS articulation to divergent political projects through which it has been mobilized as a world-making force for two centuries. But it also helps feed the insecurity and desire for certainty that we experience in conversations about our work—a desire that we have suggested is bound up with the zeitgeist of the turbulent present. This returns us to the body and to the ways in which protein gets enlisted in body projects.

Protein's status in the nutritional imagination has long been sustained by a vision of the body as a mechanistic and sovereign entity within the control of the individual subject. That sovereignty has always been a fantasy, but in the early twenty-first century the possibility of the separation of body from world is compromised like never before. Each day brings further knowledge of how our bodies are endangered by their porosity and their interdependency with an endangered, endangering planet. Microplastics in human lungs, contaminated water sources, viral diseases, and polluted air offer a short index of everyday contaminations that rupture the notion of the bounded, individuated body. As Achille Mbembe puts it, "The earth is being redefined in a way that binds all species together: humans, technology, animals, fungi, plants, viruses, bacteria—the same life in disparate bodies."[29] While other socio-ecological agents are fundamentally compromised, protein has thrived through these accelerated comminglings of bodily and environmental natures. It is even asked to provide surety and stability when vulnerability-inducing bodily enmeshments—and readily available yet contested knowledge about them—are everywhere.

This point was brought home to us as we were writing this epilogue, when we received an invitation to discuss the ecology of whey protein powder on the BBC Radio 4 series *Sliced Bread*, which "investigates the latest ad-hyped products and trending fads promising to make us healthier, happier and greener." Like hair loss products, longevity pills, DNA ancestry tests, foam rollers, and many other products before it, protein powder was to be put to the show's recurring test: "The best thing since sliced bread, or marketing BS?" With other experts called on to discuss the producers' central question about the need for and efficacy of protein supplementation, our role in the discussion was to be narrow: Could we help shed light on a listener's question about the environmental footprint of different protein powders? We offered a short audio clip on whey's history as a pollutant, the problems wrought by an industrialized food system, and protein powder as a solution that leaves that system not only intact but augmented by a new market. But our direct answer to the listener's question—that there is no powder that isn't an expression of the food system in which it is produced—was probably as unsatisfying as our "It depends" response to "How much and what kind of protein should I eat?"

Various critiques of the episode's underlying assumptions could be made here, including the nutritionist reduction of protein to its biochemical efficacy and the framing of environmental sustainability as the responsibility of the individual. But there is also something to be said for the question (another one we often hear) about which protein supplements are most environmentally friendly, given the rising ecological and geopolitical vulnerability with which people live. The listener's inquiry, imbued with an acknowledgment of humanity's ecological embeddedness and ethical care, speaks both of a desire not to perpetuate environmental harm and of a bodily practice that might ward off its encroachments under the skin. In an era characterized by multifold forms of uncertainty, it makes sense that protein is asked to offer a sense of stability in the corporeal domain. As new worlds open up and old ones slip away, the power of protein endures, for now—its fantastical promises thriving both with and against the contingencies that threaten its status.

Notes

INTRODUCTION

1. See Twine, "Emissions from Animal Agriculture."

2. MacDonald, "Producing Protein," 32. See also Adjemian et al., "Protein Politics"; Howard et al., "'Protein' Industry Convergence"; Mylan, Andrews, and Maye, "The Big Business of Sustainable Food Production and Consumption."

3. Decker, "How to Compete in the Crowded Protein Market"; and see Guthman, "Binging and Purging."

4. Decker, "How to Compete in the Crowded Protein Market."

5. See Davis and Jacobson, *Proteinaholic*.

6. See Davis and Jacobson, *Proteinaholic*.

7. See Hayes, "Measuring Protein Content in Food"; Davis and Jacobson, *Proteinaholic*; Nishimura et al., "Dietary Protein Requirements and Recommendations for Healthy Older Adults"; and Heid, "The Great Protein Debate Heats Up."

8. Institute of Medicine, *Dietary Reference Intakes*.

9. Institute of Medicine, *Dietary Reference Intakes*.

10. Institute of Medicine, *Dietary Reference Intakes*, 590.

11. See Overend, *Shifting Food Facts*.

12. See Farrington, "Take It with a Grain (or More) of Salt."

13. See Berryman et al., "Protein Intake Trends and Conformity with the Dietary Reference Intakes in the United States."

14. Davis and Jacobson, *Proteinaholic*.

15. See Belasco, *Appetite for Change*; Pollan, *In Defense of Food*; and Scrinis, *Nutritionism*.

16. See Scrinis, *Nutritionism*.

17. Rose, *The Politics of Life Itself*, 11.

18. See Cooper, *Life as Surplus*.

19. Scrinis, *Nutritionism*; Rose, "Molecular Biopolitics."

20. Martschukat, *The Age of Fitness*.

21. See Ingham, "From Public Issue to Personal Trouble."

22. See Atkinson, "Playing with Fire"; Gard and Wright, *The Obesity Epidemic*; Dworkin and Wachs, *Body Panic*; Jeffords, *Hard Bodies*; and Toffoletti and Thorpe, "Bodies, Gender, and Digital Affect."

23. See Martschukat, *The Age of Fitness*; Howell and Ingham, "From Social Problem to Personal Issue"; Ingham, "From Public Issue to Personal Trouble"; and Cole and Hribar, "Celebrity Feminism."

24. See Howell and Ingham, "From Social Problem to Personal Issue"; and Ingham, "From Public Issue to Personal Trouble."

25. See Hargreaves and Vertinsky, *Physical Culture, Power and the Body*; and Millington, *Fitness, Technology and Society*.

26. Pulidindi and Ahuja, "Protein Powder Market Size by Source."

27. Seymour, *Disaster Nationalism*, 11.

28. See Intergovernmental Panel on Climate Change, *Climate Change and Land: An IPCC Special Report on Climate Change, Desertification, Land Degradation, Sustainable Land Management, Food Security, and Greenhouse Gas Fluxes in Terrestrial Ecosystems*, 2020, http://ipcc.ch.

29. For comprehensive reviews of this literature, see Lonkila and Kaljonen, "Promises of Meat and Milk Alternatives"; and Mylan, Andrews, and Maye, "The Big Business of Sustainable Food Production and Consumption."

30. Guthman and Biltekoff, "Magical Disruption?"; Mylan, Andrews, and Maye, "The Big Business of Sustainable Food Production and Consumption."

31. Howard et al., "'Protein' Industry Convergence."

32. Guthman and Biltekoff, "Magical Disruption?"; Broad, "Making Meat, Better"; Stephens, King, and Lyall, "Blood, Meat, and Upscaling Tissue Engineering."

33. Sexton, "Food as Software."

34. Chiles, "If They Come, We Will Build It"; Sexton, Garnett, and Lorimer, "Framing the Future of Food."

35. Jönsson, "Benevolent Technotopias"; Jönsson, Linné, and McCrow-Young, "Many Meats and Many Milks?"; Lonkila and Kaljonen, "Promises of Meat and Milk Alternatives"; Stephens et al., "Bringing Cultured Meat to Market."

36. Sexton, "Eating for the Post-Anthropocene"; Van der Weele and Driessen, "Emerging Profiles for Cultured Meat."

37. Lynch and Pierrehumbert, "Climate Impacts of Cultured Meat and Beef Cattle"; Santo et al., "Considering Plant-Based Meat Substitutes"; Verbeke, Sans, and Van Loo, "Challenges and Prospects for Consumer Acceptance of Cultured Meat."

38. Guthman and Fairbairn, "Speculating on Collapse."

39. Creswell, "Beyond Meat Is Struggling."

40. Guthman and Fairbairn, "Speculating on Collapse."

41. Guthman, *The Problems with Solutions*.

42. Good Food Institute, *State of Global Policy 2023: Public Investment in Alternative Proteins to Feed a Growing World*, accessed May 12, 2025, https://gfi.org/resource/alternative-proteins-state-of-global-policy/.

43. Howard et al., "'Protein' Industry Convergence"; Guthman and Fairbairn, "Speculating on Collapse."

44. Guthman and Biltekoff ask a version of this question in their article on alternative protein development in Silicon Valley: Guthman and Biltekoff, "Agri-Food Tech's Building Block."

45. Marx, *Capital*, vol. 1.

46. See Ware, "Robbing the Soil."

47. See Appadurai, "Introduction: Commodities and the Politics of Value"; Cook, "Follow the Thing: Papaya"; Latour, *Reassembling the Social*; Marx, *Capital*; and Mol, *The Body Multiple*.

48. Appadurai, "The Thing Itself," 15.

49. See Barad, *Meeting the Universe Halfway*; and Coole and Frost, *New Materialisms*.

50. See Steffen, Crutzen, and McNeill, "The Anthropocene."

51. See Steffen, Crutzen, and McNeill, "The Anthropocene."

52. See Ahmed, "Imaginary Prohibitions"; and Davis et al., "Anthropocene, Capitalocene, . . . Plantationocene?"

53. See King and Weedon, "The Nature of the Body"; Ahmed, "Imaginary Prohibitions"; Davis et al., "Anthropocene, Capitalocene, . . . Plantationocene?"; Karera, "Blackness and the Pitfalls of Anthropocene Ethics"; and Leong, "The Mattering of Black Lives."

54. See Chakrabarty, *The Climate of History in a Planetary Age*.

55. See Baldwin and Erickson, "Introduction: Whiteness, Coloniality, and the Anthropocene"; Vergès, "Racial Capitalocene"; and Yusoff, *A Billion Black Anthropocenes*.

56. See Ahuja, *Bioinsecurities*; Davis et al., "Anthropocene, Capitalocene, . . . Plantationocene?"; and TallBear, "An Indigenous Reflection."

57. See Ahuja, *Bioinsecurities*; Davis et al., "Anthropocene, Capitalocene, . . . Plantationocene?"; TallBear, "An Indigenous Reflection"; Haraway, *When Species Meet*; and Kirksey and Helmreich, "The Emergence of Multispecies Ethnography."

58. King and Weedon, "Embodiment Is Ecological."

59. King and Weedon, "Embodiment Is Ecological."

60. Mol, "Layers or Versions?"

61. See Frost, *Biocultural Creatures*; and Myers, *Rendering Life Molecular*.

62. Myers, *Rendering Life Molecular*.

63. Sarmiento, "Umwelt, Food, and the Limits of Control," 74.

64. See Frost, *Biocultural Creatures*; and Myers, *Rendering Life Molecular*.

CHAPTER ONE. WHAT IS PROTEIN AND WHY DOES IT MATTER?

1. Glas, "The Evolution of a Scientific Concept."

2. See Carpenter, *Protein and Energy*, 48; and Liebig, "Animal Chemistry," 103.

3. See Frost, *Biocultural Creatures*; Myers, *Rendering Life Molecular*.

4. Frost, *Biocultural Creatures*, 1.

5. Frost, *Biocultural Creatures*, 27.

6. Frost, *Biocultural Creatures*, 116.

7. Myers, *Rendering Life Molecular*.

8. Myers, *Rendering Life Molecular*, 6.

9. Frost, *Biocultural Creatures*, 78.

10. Myers, *Rendering Life Molecular*, 239.

11. Myers, *Rendering Life Molecular*, 239.

12. Myers, *Rendering Life Molecular*, x.

13. Kimura, *Hidden Hunger*, 20.

14. Kimura, *Hidden Hunger*.

15. Brock, *Justus von Liebig*, ix; Browne, "Justus von Liebig," 3.

16. See Marchesi, "Justus von Liebig."

17. Brock, *Justus von Liebig*.

18. See Carpenter, "A Short History of Nutritional Science"; Vickery, "Liebig and the Chemistry of Proteins"; and Glas, "The Evolution of a Scientific Concept."

19. Vickery, "The Origin of the Word Protein," 392.

20. Carpenter, *Protein and Energy*, 43.

21. Vickery, "The Origin of the Word Protein," 388.

22. Glas, "The Evolution of a Scientific Concept," 41.

23. Vickery, "Liebig and the Chemistry of Proteins," 21.

24. Vickery, "Liebig and the Chemistry of Proteins," 21.

25. See Carpenter, "A Short History of Nutritional Science."

26. Vickery, "Liebig and the Chemistry of Proteins," 23.

27. See Vickery, "Liebig and the Chemistry of Proteins"; and Glas, "The Evolution of a Scientific Concept."

28. Glas, "The Evolution of a Scientific Concept," 44.

29. Glas, "The Evolution of a Scientific Concept"; and see Tanford and Reynolds, *Nature's Robots*.

30. Tanford and Reynolds, *Nature's Robots*.

31. Tanford and Reynolds, *Nature's Robots*, 24.

32. Tanford and Reynolds, *Nature's Robots*, 23.

33. Tanford and Reynolds, *Nature's Robots*, 27.

34. Glas, "The Evolution of a Scientific Concept," 44.

35. Glas, "The Evolution of a Scientific Concept," 45.

36. Tanford and Reynolds, *Nature's Robots*, 18.

37. Proceedings of the Physiological Society, "Protein Nomenclature," xvii.

38. Proceedings of the Physiological Society, "Protein Nomenclature," xvii.

39. Proceedings of the Physiological Society, "Protein Nomenclature," xvii–xviii.

40. Proceedings of the Physiological Society, "Protein Nomenclature," xviii.

41. Proceedings of the Physiological Society, "Protein Nomenclature," xviii.

42. See Glas, "The Evolution of a Scientific Concept."

43. Vickery and Schmidt, "History of the Discovery of the Amino Acids," 171.

44. Howes, "Many of Our Proteins Remain Hidden in the Dark Proteome."

45. Mol, *The Body Multiple*.

46. Ahlgren, "What Is a Protein?"

47. Garrett et al., *Biochemistry*, 17.

48. Frost, *Biocultural Creatures*, 79.

49. Havstad, "Protein Tokens, Types, and Taxa," 75.

50. Myers, *Rendering Life Molecular*, 242.

51. Myers, *Rendering Life Molecular*, 242.

52. Havstad, "Protein Tokens, Types, and Taxa," 77.

53. Havstad, "Protein Tokens, Types, and Taxa," 75.

54. Garrett et al., *Biochemistry*, 133.

55. Jacobsen, "Artificial Proteins."

56. See Service, "'The Game Has Changed.'"

57. Hesman-Saey, "Has AlphaFold Actually Solved Biology's Protein-Folding Problem?"

58. Glas, "The Evolution of a Scientific Concept," 51.

59. Glas, "The Evolution of a Scientific Concept," 51.

60. Glas, "The Evolution of a Scientific Concept," 51.

61. Glas, "The Evolution of a Scientific Concept," 52.

62. See Carpenter, "The History of Enthusiasm for Protein," 1364.

63. Carpenter, "The History of Enthusiasm for Protein," 1365.

64. Carpenter, *Protein and Energy*, 33, 35.

65. See Moulton, *Liebig and After Liebig*; and Brock, *Justus von Liebig*.

66. Carpenter, "A Short History of Nutritional Science," 641.

67. Liebig, *Researches on the Chemistry of Food*, 45.

68. See Suzanne Girard Eberle, "The Body's Fuel Sources," Human Kinetics, accessed June 7, 2023, https://us.humankinetics.com/blogs/excerpt/the-bodys-fuel-sources.

69. Carpenter, "The History of Enthusiasm for Protein," 1365.

70. Carpenter, *Protein and Energy*.

71. Herzog, "Skeletal Muscle Mechanics," 14.

72. Carpenter, "The History of Enthusiasm for Protein," 1365.

73. Kuhn, *The Structure of Scientific Revolutions*.

74. See Marchesi, "Justus von Liebig."

75. See Browne, "Justus von Liebig"; Marchesi, "Justus von Liebig"; and Munday, "Politics by Other Means."

76. See Brock, *Justus von Liebig*; Carpenter, *Protein and Energy*; and Vickery, "Liebig and the Chemistry of Proteins."

77. See Brock, *Justus von Liebig*; Marchesi, "Justus von Liebig"; and Munday, "Politics by Other Means."

78. See Brock, *Justus von Liebig*; and Munday, "Politics by Other Means."

79. Marchesi, "Justus von Liebig," 212. See also Brock, *Justus von Liebig*.

80. Liebig, "The Nutritive Value of Different Sorts of Food," 4.

81. Liebig, "The Nutritive Value of Different Sorts of Food," 4, 5, 186.

82. Liebig, "The Nutritive Value of Different Sorts of Food," 4, 36.

83. Liebig, "The Nutritive Value of Different Sorts of Food," 5.

84. Liebig, "The Nutritive Value of Different Sorts of Food," 37.

85. Liebig, "The Nutritive Value of Different Sorts of Food," 37.

86. Liebig, "The Nutritive Value of Different Sorts of Food," 37.

87. Liebig, "The Nutritive Value of Different Sorts of Food," 114.

88. Liebig, "The Nutritive Value of Different Sorts of Food," 357.

89. Liebig, "The Nutritive Value of Different Sorts of Food," 357, 358.

90. Liebig, "The Nutritive Value of Different Sorts of Food," 358.

91. See Liebig, "Animal Chemistry," 48.

92. Marchesi, "Justus von Liebig," 206.

93. Marchesi, "Justus von Liebig," 213.

94. Marchesi, "Justus von Liebig," 213.

95. Marchesi, "Justus von Liebig," 207.

96. Warren, *Meat Makes People Powerful*, 1.

97. Warren, *Meat Makes People Powerful*, 2.

98. Brock, *Justus von Liebig*, 228.

99. Finlay, "Early Marketing of the Theory of Nutrition," 49.

100. Brock, *Justus von Liebig*, 224. See also Lewowicz, LEMCO.

101. See Lewowicz, LEMCO; Sunseri, "International Beef Packing in the Age of Empire."

102. See de Rennie, "Advertising Capitalism."

103. Finlay, "Quackery and Cookery," 404.

104. Finlay, "Early Marketing of the Theory of Nutrition," 48.

105. See Roy, *Alimentary Tracts*.

106. Gandhi, *Affective Communities*, 82.

107. Berson, *The Meat Question*.

108. See Carpenter, "A Short History of Nutritional Science."

109. See Carpenter, "The History of Enthusiasm for Protein."

110. Carpenter, "The History of Enthusiasm for Protein," 1366.

111. Berson, "Race and the Science of Starvation."

112. Berson, "Race and the Science of Starvation."

113. Carpenter, *Protein and Energy*, 117.

114. Gandhi, *Affective Communities*, 81–82.

115. Roy, *Alimentary Tracts*, 8.

116. Roy, *Alimentary Tracts*, 78.

117. See Gandhi, *Affective Communities*.

118. See Gandhi, *Affective Communities*.

119. Carpenter, *Protein and Energy*, 114.

120. Carpenter, *Protein and Energy*, 140.

121. Mendel, *Nutrition*, 14, 110.

CHAPTER TWO. THE GREAT PROTEIN FIASCO, THEN AND NOW

1. McLaren, "The Great Protein Fiasco."

2. McLaren, "The Great Protein Fiasco"; Konotey-Ahulu and McLaren, "Issues in Kwashiorkor"; Nott, "'How Little Progress'?"

3. Nott, "'How Little Progress'?"; McLaren, "A Fresh Look at Protein-Calorie Malnutrition."

4. See Carpenter, *Protein and Energy*; Kimura, *Hidden Hunger*; Nott, "'How Little Progress'?"; Nott, "'No One May Starve in the British Empire.'"

5. Nott, "'How Little Progress'?"; Nott, "'No One May Starve in the British Empire.'"

6. Nott, "'How Little Progress'?"; Nott, "'No One May Starve in the British Empire.'"

7. Scott-Smith, *On an Empty Stomach*.

8. Scott-Smith, *On an Empty Stomach*, 124.

9. Carpenter, *Protein and Energy*.

10. Kimura, *Hidden Hunger*; Nott, "'How Little Progress'?"

11. Carpenter, *Protein and Energy*, xi. The two earlier periods Carpenter mentions here are discussed in chapter 1 and refer to challenges to Liebig's protein-first philosophy in the mid-nineteenth century and to the Chittenden-McCay debate at the start of the twentieth century, respectively.

12. Carpenter, *Protein and Energy*.

13. Carpenter, "The History of Enthusiasm for Protein."

14. Kimura, *Hidden Hunger*.

15. Cannon, "Nutrition."

16. Cannon, "Nutrition," S484.

17. Cannon, "The Rise and Fall of Dietetics and of Nutrition Science."

18. Carpenter, *Protein and Energy*; Nott, "'How Little Progress'?"; Scott-Smith, "The Fetishism of Humanitarian Objects."

19. Nkhoma, "'We Are What We Eat'"; Nkhoma, "The 'Malnutrition Syndrome'"; Vaughan, "Food Production and Family Labor in Southern Malawi"; Vaughan, *The Story of an African Famine*.

20. Smith, "The Emergence of Vitamins as Bio-Political Objects."

21. "An overtly political concern": Nott, "'No One May Starve in the British Empire,'" 558. "Biopolitical objects": Smith, "The Emergence of Vitamins as Bio-Political Objects."

22. Kimura, *Hidden Hunger*.

23. Worboys, "The Discovery of Colonial Malnutrition."

24. Nott, "'No One May Starve in the British Empire,'" 559.

25. Nott, "'No One May Starve in the British Empire,'" 559.

26. Cannon, "Nutrition," S486.

27. Gilks and Orr, "The Nutritional Condition of the East African Native"; Orr and Gilks, *Studies of Nutrition*.

28. Worboys, "The Discovery of Colonial Malnutrition."

29. Gilks and Orr, "The Nutritional Condition of the East African Native," 560; Blaxter and Garnett, *Primed for Power*.

30. Brantley, "Kikuyu-Maasai Nutrition and Colonial Science," 85.

31. Brantley, "Kikuyu-Maasai Nutrition and Colonial Science," 85.

32. Gilks and Orr, "The Nutritional Condition of the East African Native."

33. Nott, "'No One May Starve in the British Empire,'" 559.

34. Brantley, "Kikuyu-Maasai Nutrition and Colonial Science."

35. Fulton, "Survey of Meat Supplies and Distribution in the Gold Coast," cited in Nott, "'No One May Starve in the British Empire,'" 559.

36. Kimura, *Hidden Hunger*; Worboys, "The Discovery of Colonial Malnutrition."

37. Williams, "A Nutritional Disease of Childhood"; Williams, "Kwashiorkor"; Semba, "The Rise and Fall of Protein Malnutrition."

38. Craddock, *Retired, Except on Demand*.

39. Williams, "A Nutritional Disease of Childhood."

40. Williams, "A Nutritional Disease of Childhood," 432.

41. Carpenter, *Protein and Energy*, 142–43.

42. Carpenter, *Protein and Energy*, 143.

43. Carpenter, *Protein and Energy*, 143.

44. Nott, "'No One May Starve in the British Empire,'" 555.

45. Nott, "'No One May Starve in the British Empire,'" 565–66.

46. Nott, "'No One May Starve in the British Empire,'" 566.

47. Williams, "A Nutritional Disease of Childhood"; Williams, "Kwashiorkor."

48. Kimura, *Hidden Hunger*.

49. Gordon and Hobbes, "The Great Protein Fiasco"; Stanton, "Listening to the Ga," 158.

50. Gordon and Hobbes, "The Great Protein Fiasco."

51. Nott, "'No One May Starve in the British Empire,'" 570; Stanton, "Listening to the Ga."

52. Konotey-Ahulu and McLaren, "Issues in Kwashiorkor," 548.

53. Konotey-Ahulu, "There Is Nothing Mysterious About Kwashiorkor."

54. BMJ Best Practice, "Kwashiorkor"; Pencharz et al., "Current Issues in Determining Dietary Protein and Amino-Acid Requirements."

55. In Nott, "'No One May Starve in the British Empire.'"

56. Nott, "'No One May Starve in the British Empire,'" 554.

57. Nott, "'No One May Starve in the British Empire,'" 555; Gordon and Hobbes, "The Great Protein Fiasco."

58. Nott, "'No One May Starve in the British Empire.'"

59. Nott, "'No One May Starve in the British Empire,'" 567.

60. Nott, "'No One May Starve in the British Empire,'" 567.

61. Nott, "'No One May Starve in the British Empire.'"

62. Nott, "'No One May Starve in the British Empire.'"

63. Nott, "'No One May Starve in the British Empire,'" 568.

64. Williams, "Kwashiorkor," 1151.

65. Nott, "'No One May Starve in the British Empire,'" 572.

66. Nott, "'No One May Starve in the British Empire,'" 571.

67. Nott, "'No One May Starve in the British Empire,'" 574.

68. Trowell, "Malignant Malnutrition," 426; Trowell and Davis, "Kwashiorkor."

69. Trowell and Davis, "Kwashiorkor," 796.

70. Brock and Autret, *Kwashiorkor in Africa*, 66.

71. Brock and Autret, *Kwashiorkor in Africa*, 32–33.

72. Nott, "'No One May Starve in the British Empire,'" 574.

73. Gordon and Hobbes, "The Great Protein Fiasco."

74. Brock and Autret, *Kwashiorkor in Africa*, 7.

75. Carpenter, *Protein and Energy*, 149.

76. McLaren, "The Great Protein Fiasco," 94.

77. Cannon, "Nutrition," S485.

78. Brock and Autret, *Kwashiorkor in Africa*, 45–46.

79. Brock and Autret, *Kwashiorkor in Africa*, 65.

80. Carpenter, *Protein and Energy*, 149.

81. McLaren, "The Great Protein Fiasco," 93.

82. McLaren, "The Great Protein Fiasco."

83. Carpenter, *Protein and Energy*, 156.

84. Carpenter, *Protein and Energy*, 156.
85. Carpenter, *Protein and Energy*, 157.
86. Carpenter, *Protein and Energy*, 157.
87. Black, *The Children and the Nations*, 146.
88. Scott-Smith, *On an Empty Stomach*, 131.
89. Scott-Smith, *On an Empty Stomach*.
90. Semba, "The Rise and Fall of Protein Malnutrition."
91. Carpenter, *Protein and Energy*.
92. Carpenter, *Protein and Energy*, 158.
93. McMichael, "A Food Regime Genealogy."
94. Carpenter, *Protein and Energy*, 158.
95. Carpenter, *Protein and Energy*, 158.
96. Carpenter, *Protein and Energy*, 159.
97. Carpenter, *Protein and Energy*, 159.
98. See Semba, "The Rise and Fall of Protein Malnutrition."
99. See Semba, "The Rise and Fall of Protein Malnutrition."
100. Semba, "The Rise and Fall of Protein Malnutrition," 81.
101. Scott-Smith, *On an Empty Stomach*, 142.
102. Food and Agriculture Organization of the United Nations, *Lives in Peril*.
103. McLaren, "The Great Protein Fiasco."
104. Semba, "The Rise and Fall of Protein Malnutrition."
105. Carpenter, *Protein and Energy*, 162.
106. Nott, "'How Little Progress'?"; Carpenter, *Protein and Energy*.
107. Nott, "'How Little Progress'?," 774.
108. Nott, "'How Little Progress'?," 774.
109. McLaren, "The Great Protein Fiasco," 93.
110. Konotey-Ahulu, "There Is Nothing Mysterious About Kwashiorkor."
111. Nott, "'How Little Progress'?," 774.
112. Sathyamala, "Nutritionalizing Food," 825.
113. McLaren, "The Great Protein Fiasco Revisited."
114. Sai, "Nutrition in Ghana," 100; Nott, "'How Little Progress'?," 774.
115. Jelliffe, "Commerciogenic Malnutrition?"
116. Scott-Smith, *On an Empty Stomach*.
117. McLaren, "The Great Protein Fiasco"; McLaren, "The Great Protein Fiasco Revisited."
118. Scott-Smith, *On an Empty Stomach*.
119. McLaren, "The Great Protein Fiasco," 96.
120. Nott, "'How Little Progress'?"
121. McLaren, "The Great Protein Fiasco," 95.
122. Semba, "The Rise and Fall of Protein Malnutrition," 84.
123. Semba, "The Rise and Fall of Protein Malnutrition," 84.
124. Scott-Smith, *On an Empty Stomach*, 152; Brock and Hansen, "Protein Requirement."
125. Semba, "The Rise and Fall of Protein Malnutrition."
126. McLaren, "The Great Protein Fiasco."

127. Rivers, "The Profession of Nutrition."

128. Semba, "The Rise and Fall of Protein Malnutrition," 86.

129. Semba, "The Rise and Fall of Protein Malnutrition"; Kimura, *Hidden Hunger*.

130. Semba, "The Rise and Fall of Protein Malnutrition."

131. Semba, "The Rise and Fall of Protein Malnutrition," 86.

132. Semba, "The Rise and Fall of Protein Malnutrition," 86.

133. Nott, "'How Little Progress'?"

134. Scott-Smith, *On an Empty Stomach*, 164.

135. Carpenter, *Protein and Energy*.

136. Lappe, *Diet for a Small Planet*.

137. Carpenter, *Protein and Energy*.

CHAPTER THREE. FROM GUTTER TO GOLD

Chapter 3 is a revised version of work originally published as "Embodiment Is Ecological: The Metabolic Lives of Whey Protein Powder," *Body and Society* 26, no. 1 (2020): 82–106. Material in chapter 3 also appeared in "The Nature of the Body in Sport and Physical Culture: From Bodies and Environments to Ecological Embodiment," *Sociology of Sport Journal* 38, no. 2 (2021): 131–39.

1. See Tunick, "Whey Protein Production and Utilization."

2. See Jhally, *The Codes of Advertising*.

3. Scanlon, *On Garbage*, 9.

4. Douglas, *Purity and Danger*, 165.

5. Latour, *We Have Never Been Modern*, 10.

6. Blanchette, "Living Waste," 91; see also Blanchette, *Porkopolis*.

7. Blanchette, "Living Waste," 82.

8. "Arbiters of each other's growth": Blanchette, "Living Waste," 88.

9. "Whey Protein Market Size"; Smithers, "Whey and Whey Proteins."

10. See Wiley, *Cultures of Milk*.

11. See Luiz, "My Beef with Dairy."

12. Moss, "While Warning About Fat."

13. Bentley, "Trends in U.S. Per Capita Consumption of Dairy Products."

14. Paxson, *The Life of Cheese*.

15. Jelen, "Whey Processing." For the "nine liters" figure, see Onwulata and Huth, *Whey Processing, Functionality and Health Benefits*.

16. United States Department of Agriculture, "Whey to Ethanol."

17. Dewey, "America's Cheese Stockpile."

18. Landecker, "A Metabolic History of Manufacturing Waste," 531.

19. Lougheed, "Actor Network Theory Examination."

20. Smithers, "Whey and Whey Proteins."

21. See Crowfoot and Wondolleck, *Environmental Disputes*; Knight, "Whey Disposal Troubles Vermonters"; Lougheed, "Actor-Network Theory Examination"; Menzies, *By the Labor of Their Hands*; and Scott, "Domtar Paper Mill."

22. See Borre, "Big Data Arrives."

23. For the responses of government, see Bertin, "Ontario Program"; of cheese-makers, see Knight, "Whey Disposal Troubles Vermonters"; and of food industry researchers, see Immen, "Human-Life Cow's Milk."

24. Scrinis, *Nutritionism*.

25. Smithers, "Whey and Whey Proteins," 696.

26. Mbembe, "Futures of Life."

27. Rose, *The Politics of Life Itself*.

28. Marx, *Capital*, vol. 3, 949.

29. Moore, *Capitalism in the Web of Life*.

30. Moore, *Capitalism in the Web of Life*, 2.

31. Landecker, "A Metabolic History of Manufacturing Waste," 168.

32. See Cooper, *Life as Surplus*; and Sunder Rajan, *Biocapital*.

33. See King and Weedon, "The Nature of the Body."

34. See Guthman, "Binging and Purging."

35. Smithers, "Whey and Whey Proteins."

36. Adams, *The Sexual Politics of Meat*.

37. Mulder, "The Quest for Sustainable Nitrogen Removal Technologies."

38. Ledin, "The Protein Interview."

39. Erisman et al., "Nitrogen," 3.

40. Erisman et al., "Nitrogen," 3.

41. Page, "'The Greatest Victory,'" 385.

42. Westgate and Park, "Evaluation of Proteins," 5352.

43. See Moran, "Wastewater Is Key to Reducing Nitrogen Pollution"; and Urbańczyk, Sowa, and Simka, "Urea Removal."

44. See Liboiron, "Redefining Pollution and Action"; and Murphy, *The Economization of Life*.

45. See Shotwell, *Against Purity*.

CHAPTER FOUR. A POVERTY OF FLESH?

1. See Heffernan, "Physical Culture Supplements"; and Haushofer, "Between Food and Medicine."

2. "Elderly Nutrition Market Analysis," *Coherent Market Insights*, August 2022, https://www.coherentmarketinsights.com/market-insight/elderly-nutrition-market-2437.

3. "Senior Nutrition."

4. See Schöley et al., "Life Expectancy Changes Since COVID-19"; and "Global Issues: Ageing," United Nations, accessed June 7, 2023, https://www.un.org/en/global-issues/ageing.

5. Hacking, *Historical Ontology*.

6. Rosenberg, "Sarcopenia."

7. Rosenberg, "Sarcopenia."

8. Rosenberg, "Sarcopenia" (2011), 337.

9. See Grosz, *Volatile Bodies*; and Alcoff, "Philosophy Matters."

10. Foucault, *The Birth of the Clinic*, 232.

11. Evans, "What Is Sarcopenia?"

12. See Clark and Manini, "What Is Dynapenia?"

13. Rosenberg, "Sarcopenia."

14. Drew, "Lifting the Burden," S16.

15. Drew, "Lifting the Burden," S16.

16. Marcell, "Sarcopenia," M912.

17. Avgerinou, "Sarcopenia," 200.

18. Marcell, "Sarcopenia," M911.

19. Cruz-Jentoft, Baeyens, et al., "Sarcopenia."

20. Chumlea et al., "International Working Group on Sarcopenia"; Chen et al., "Sarcopenia in Asia."

21. Gustafsson and Ulfhake, "Sarcopenia."

22. Haase, Brodersen, and Bülow, "Sarcopenia."

23. See Truog et al., "Brain Death at Fifty."

24. See Howell and Ingham, "From Social Problem to Personal Issue"; and Ingham, "From Public Issue to Personal Trouble."

25. Martschukat, *The Age of Fitness*, 3; Foucault, *Discipline and Punish*.

26. See American College of Sports Medicine, "General Principles of Exercise Prescription"; Campbell, De Jesus, and Prapavessis, "Physical Fitness."

27. Crawford, "Healthism and the Medicalization of Everyday Life," 365. See also Crawford, "Health as a Meaningful Social Practice."

28. Metzl and Kirkland, *Against Health*.

29. Metzl and Kirkland, *Against Health*, 1–2.

30. Crawford, "Healthism and the Medicalization of Everyday Life."

31. Crawford, "Health as a Meaningful Social Practice," 402

32. Crawford, "Health as a Meaningful Social Practice," 407.

33. See Schrecker, "Neoliberalism and Health."

34. Crawford, "Health as a Meaningful Social Practice," 407–8.

35. Crawford, "Health as a Meaningful Social Practice," 409.

36. Crawford, "Health as a Meaningful Social Practice," 410.

37. Rose, *The Politics of Life Itself*, 16.

38. Crary, *Scorched Earth*, 67–68.

39. Crary, *Scorched Earth*, 68.

40. Schrecker, "Neoliberalism and Health."

41. See Murphy, *The Economization of Life*; and Patchin, "For the Sake of the Child."

42. Vespa, "The U.S. Joins Other Countries."

43. Vespa, "The U.S. Joins Other Countries."

44. "Infographic: Canada's Seniors Population Outlook."

45. Cooper, "A Burden on Future Generations?," 744.

46. Shalal, "Aging Population to Hit U.S. Economy."

47. Kenny, "The Biopolitics of Global Health," 21.

48. "The Pensions Act 2007," UK National Archives, last modified October 3, 2012, https://webarchive.nationalarchives.gov.uk/ukgwa/20130125095415/http://www.dwp.gov.uk/policy/pensions-reform/the-pensions-act-2007/.

49. Marcell, "Sarcopenia," M911.

50. Gustafsson and Ulfhake, "Sarcopenia."

51. Cruz-Jentoft, Bahat, et al., "Sarcopenia," 17.

52. Goates et al., "Economic Impact of Hospitalizations."

53. See Kenny, "The Biopolitics of Global Health"; Murphy, *The Economization of Life*; Patchin, "For the Sake of the Child."

54. Goates et al., "Economic Impact of Hospitalizations," 98.

55. Janssen et al., "The Healthcare Costs of Sarcopenia."

56. Evans, "What Is Sarcopenia?"

57. Anker, Morley, and von Haehling, "Welcome to the ICD-10 Code for Sarcopenia."

58. Poulson, Trevino, and Sather, "Leucine."

59. "Senior Nutrition."

60. Haigh, "Protein Boost for Boost."

61. Dunn, "Why Grandma's Nutrition Drink."

62. Biophytis, "Sarconeos (BIO101)," accessed June 5, 2023, https://www.biophytis.com/en/our-science/sarcopenie/

63. Vince, *Nomad Century*, xiii.

64. Vince, *Nomad Century*.

65. Moore, *Capitalism in the Web of Life*.

CHAPTER FIVE. PROTEIN IN THE MUSCULAR MANOSPHERE

1. See Pulidindi and Ahuja, "Protein Powder Market Size by Source."

2. Creatine is usually lab-generated from plant-based compounds.

3. See Atkinson, "Playing with Fire."

4. See Gago and Malo, "Introduction: The New Feminist Internationale."

5. See Bratich, *On Microfascism*.

6. Marcus, *Own the Day, Own Your Life*.

7. Anderson, "Onnit to Be Acquired by Unilever."

8. Marcus, *Own the Day, Own Your Life*.

9. Marcus, "The Sacred Masculine Archetype"; Marcus, *Own the Day, Own Your Life*.

10. Homayun, "How Aubrey Marcus Built Onnit into a $28-Million Company in 5 Years."

11. Atkinson, "Playing with Fire," 179.

12. Atkinson, "Playing with Fire," 182.

13. See Messner, *Politics of Masculinities*.

14. See Connell, *Masculinities*.

15. See Jane, *Misogyny Online*.

16. Kelly, DiBranco, and DeCook, "Misogynist Incels and Male Supremacism," 18.

17. See Kusz and Hodler, "'Saturdays Are for the Boys'"; Berlant, "Big Man"; Bratich, *On Microfascism*; Wenger, "Liver King and the Problematic Rise of Primal Manhood"; Gleig and Artinger, "#BuddhistCultureWars"; Reisenwitz, "Do You Even Trip, Bro?"; Spencer, "'Eating Clean' for a Violent Body"; Sugiura, *The Incel Rebellion*; Worthen, "The Podcast Bros Want to Optimize Your Life"; Pearson, "Inside the World of 'Bitcoin Carnivores'"; Ward and Voas, "The Emergence of Conspirituality"; Ging, "Alphas, Betas, and Incels"; and see Wyman, "Bro Culture."

18. Ashcraft, "Andrew Tate's Manosphere Is No Laughing Matter."

19. Banet-Weiser, *Empowered*, 37.

20. Gibson-Graham, *A Postcapitalist Politics*, 38.

21. Bratich, *On Microfascism*.

22. Bratich, *On Microfascism*, 23. See also Cavallero and Gago, "Feminism, the Pandemic, and What Comes Next"; Deleuze and Guattari, *A Thousand Plateaus*; Gago and Malo, "Introduction: The New Feminist Internationale."

23. Bratich, *On Microfascism*, 5.

24. Bratich, *On Microfascism*, 8.

25. Bratich, *On Microfascism*, 49.

26. Harnish, "How to Build a Better Bro."

27. Harnish, "How to Build a Better Bro."

28. Bratich, *On Microfascism*, 13.

29. Bratich, *On Microfascism*, 21.

30. Bratich, *On Microfascism*, 28

31. "What's Your Discipline?," JOCKO Fuel, accessed May 30, 2023, https://store.jockofuel.com/whats-your-discipline/.

32. Willink and Babin, *Extreme Ownership*.

33. Wyman, "Bro Culture." See also, McGuire-Adams, *Indigenous Feminist Gikendaasowin (Knowledge)*.

34. Wyman, "Bro Culture."

35. Wyman, "Bro Culture."

36. Wyman, "Bro Culture."

37. Wyman, "Bro Culture."

38. Wyman, "Bro Culture."

39. Wyman, "Bro Culture."

40. Wyman, "Bro Culture."

41. Remski, "Bro Science Manifesto."

42. See Onnit, "Nutrition," accessed August 31, 2024, https://www.onnit.com/.

43. See Mansal Denton, "Sacred Hunting Is a Rite of Passage," accessed August 31, 2024, https://www.sacredhunting.com/.

44. See J. P. Sears, "Sponsors," *Awaken with JP*, accessed February 1, 2023, https://awakenwithjp.com/pages/sponsors.

45. See Brown, "Alex Jones's Media Empire."

46. See Brown, "Alex Jones's Media Empire."

47. Illiberalism.org, "Jack Z. Bratich on Microfascism," June 23, 2022, https://www.illiberalism.org/jack-z-bratich-on-microfascism/.

48. Illiberalism.org, "Jack Z. Bratich on Microfascism."

49. Illiberalism.org, "Jack Z. Bratich on Microfascism."

50. Bratich, *On Microfascism*, 17, 12.

51. Bratich, *On Microfascism*, 12.

52. Bratich, *On Microfascism*, 64.

53. Bratich, *On Microfascism*, 64.

54. Peterson, *Beyond Order*.

55. Individuation Portal, "Aubrey Marcus and Joe Rogan."

56. Remski, "Bro Science Manifesto."

57. Martschukat, *The Age of Fitness*, 76.

58. See Albelda, Bell-Pasht, and Konstantinidis, "Gender and Precarious Work in the United States."

59. See TheRealYoungJamie, "Jocko Willink Blasts COVID 'Experts' Lockdown Insistence," Reddit, accessed May 30, 2023, https://www.reddit.com/r/JoeRogan/comments /mwg2u4/jocko_willink_blasts_covid_experts_lockdown/.

60. Nat Eliason, "*Own the Day Own Your Life* by Aubrey Marcus," accessed May 30, 2023, https://www.nateliason.com/notes/own-the-day-own-your-life-by-aubrey-marcus.

61. Fox News, "Jocko Willink Blasts COVID 'Experts' Lockdown Insistence."

62. Remski, "Bro Science Manifesto."

63. Bratich and Banet-Weiser, "From Pick-Up Artists to Incels," 5008.

64. Oppenheim, "Figures That Lay Bare the Shocking Scale."

EPILOGUE

1. Finlay, "Quackery and Cookery," 404.

2. Technavio, *High Protein-Based Food Market Analysis: North America, Europe, APAC, South America, Middle East and Africa, US, China, France, Germany, Russia—Size and Forecast 2024–2028*, 2024, https://www.technavio.com/report/high-protein-based-food -market-industry-analysis#:~:text=The%20high%20protein%2Dbased%20food,due%20 to%20several%20key%20drivers.

3. OECD-FAO, "Agricultural Outlook 2023–2032," 2023, https://www.oecd.org/en /publications/2023/07/oecd-fao-agricultural-outlook-2023-2032_859ba0c2.html.

4. Clinique, "Ingredient to Know: Whey Protein," https://www.clinique.ca/skin-school -blog/ingredient-to-know/whey-protein?srsltid=AfmBOoqKJuekKFozV44MGguj2zGL7g D3kAiPk-Gsn00oI9LFaR3oJz7w.

5. For accessible and high-level overviews of these debates, see "When It Comes to Protein," Harvard Health Publishing, July 23, 2024, https://www.health.harvard.edu /nutrition/when-it-comes-to-protein-how-much-is-too-much; Heid, "The Great Protein Debate Heats Up"; IPES (International Panel of Experts on Sustainable Food Systems), *The Politics of Protein: Examining Claims About Livestock, Fish, Alternative Proteins and Sustainability*, 2022, https://ipes-food.org/report/the-politics-of-protein/.

6. IPES, *The Politics of Protein*; Semba, "The Rise and Fall of Protein Malnutrition."

7. Office of Disease Prevention and Health Promotion, *White House Conference on Hunger, Nutrition, and Health*, 2022, https://odphp.health.gov/our-work/nutrition -physical-activity/white-house-conference-hunger-nutrition-and-health.

8. Wolfe et al., "Optimizing Protein Intake."

9. Wolfe et al., "Optimizing Protein Intake."

10. Overend, *Shifting Food Facts*, 5.

11. Overend, *Shifting Food Facts*, 2.

12. Heldke, 'It's Chomping," 252.

13. See Seymour, *The Twittering Machine*.

14. See Creswell, "Beyond Meat Is Struggling"; Jingnan, "How a Conspiracy Theory About Eating Bugs"; and Dutkiewicz and Rosenberg, "Why Right Wingers Are So Afraid."

15. See Gambert and Linné, "From Rice Eaters to Soy Boys."

16. See Hosie, "Soy Boy."

17. Temple, "Bill Gates."

18. See Jingnan, "How a Conspiracy Theory About Eating Bugs."

19. Purdy, "Functional Foods Are Boring."

20. Dutkiewicz and Rosenberg, "Why Right Wingers Are So Afraid."

21. Benson, "The War Against Beef."

22. See Creswell, "Beyond Meat Is Struggling"; Osaka, "The Big Problem with Plant-Based Meat"; and Scott-Reid, "Opinion: 'Ultraprocessed' Plant-Based Meat."

23. Dutkiewicz and Rosenberg, "Why Right Wingers Are So Afraid."

24. Dutkiewicz and Rosenberg, "Why Right Wingers Are So Afraid."

25. Al Jazeera, "How Controversial Is Trump's Pick of RFK?"

26. This movement has been longer in the making than much of the recent commentary recognizes. See Turner, *From Counterculture to Cyberculture*, for an account of how Silicon Valley's countercultural roots and liberal techno-optimism has converged with an ethos of free-market entrepreneurialism, as well as Barbrook and Cameron, "The Californian Ideology."

27. Howard et al., "'Protein' Industry Convergence."

28. Howard et al., "'Protein' Industry Convergence."

29. Mbembe, "Futures of Life," 17.

Bibliography

Adams, Carol J. *The Sexual Politics of Meat: A Feminist-Vegetarian Critical Theory.* 20th anniversary ed. London: Bloomsbury Academic, 2020.

Adjemian, Maro, Heidi Janes, Sarah J. Martin, Charles Mather, and Madelyn White. "Protein Politics: Sustainable Protein and the Logic of Energy." *Canadian Food Studies* 11, no. 1 (2024): 47–65. https://doi.org/10.15353/cfs-rcea.v11i1.628.

Ahlgren, Nathan. "What Is a Protein? A Biologist Explains." *The Conversation,* January 13, 2021. https://theconversation.com/what-is-a-protein-a-biologist-explains -152870#:~:text=Scientists%20are%20not%20exactly%20sure,causing%20your%20 muscles%20to%20work.

Ahmed, Sara. "Imaginary Prohibitions: Some Preliminary Remarks on the Founding Gestures of the 'New Materialism.'" *European Journal of Women's Studies* 15, no. 1 (2008): 23–39. https://doi.org/10.1177/1350506807084854.

Ahuja, Neel. *Bioinsecurities: Disease Interventions, Empire, and the Government of Species.* Durham, NC: Duke University Press, 2016.

Albelda, Randy, Aimee Bell-Pasht, and Charalampos Konstantinidis. "Gender and Precarious Work in the United States: Evidence from the Contingent Work Supplement 1995–2017." *Review of Radical Political Economics* 52, no. 3 (2020): 542–63. https://doi.org/10.1177/0486613419891175.

Alcoff, Linda Martin. "Philosophy Matters: A Review of Recent Work on Feminist Philosophy." *Signs: Journal of Women in Culture and Society* 25, no. 3 (2000): 841–42. https://doi.org/10.1086/495484.

Al Jazeera. "How Controversial Is Trump's Pick of RFK as US Health Secretary?" November 15, 2024. https://www.aljazeera.com/news/2024/11/15/how-controversial-is -trumps-pick-of-rfk-jr-as-us-health-secretary.

American College of Sports Medicine. "General Principles of Exercise Prescription." In *Exercise Testing and Prescription,* edited by Madeline Paternostro Bayles and Ann M. Swank. New York: Wolters Kluwer, 2018. https://www.acsm.org/docs/default-source /publications-files/acsms-exercise-testing-prescription.pdf?sfvrsn=111e9306_4.

Anderson, Will. "Onnit to Be Acquired by Unilever." *Austin Business Journal,* April 26, 2001. https://www.bizjournals.com/austin/news/2021/04/26/m-a-wrap-onnit-firstclose .html.

Anker, Stefan D., John E. Morley, and Stephan von Haehling. "Welcome to the ICD-10 Code for Sarcopenia." *Journal of Cachexia, Sarcopenia and Muscle* 7, no. 5 (2016): 512–14. https://doi.org/10.1002/jcsm.12147.

Appadurai, Arjun. "Introduction: Commodities and the Politics of Value." In *The Social Life of Things: Commodities in Cultural Perspective*, edited by Arjun Appadurai. Cambridge: Cambridge University Press, 1986.

Appadurai, Arjun. "The Thing Itself." *Public Culture* 18, no. 1 (2006): 15–22. https://doi.org/10.1215/08992363-18-1-15.

Ashcraft, Karen L. "Andrew Tate's Manosphere Is No Laughing Matter." *Transforming Society* (blog), March 16 2023. https://www.transformingsociety.co.uk/2023/03/16/andrew-tates-manosphere-is-no-laughing-matter/.

Atkinson, Michael. "Playing with Fire: Masculinity, Health, and Sports Supplements." *Sociology of Sport Journal* 24, no. 2 (2007): 165–86. https://doi.org/10.1123/ssj.24.2.165.

Avgerinou, Christina. "Sarcopenia: Why It Matters in General Practice." *British Journal of General Practice* 70, no. 693 (2020): 200–201. https://doi.org/10.3399/bjgp20X709253.

Baldwin, Andrew, and Bruce Erickson. "Introduction: Whiteness, Coloniality, and the Anthropocene." *Environment and Planning D: Society and Space*, no. 38 (2020): 3–11.

Banet-Weiser, Sarah. *Empowered: Popular Feminism and Popular Misogyny*. Durham, NC: Duke University Press, 2018. https://doi.org/10.1215/9781478002772.

Barad, Karen. *Meeting the Universe Halfway: Quantum Physics and the Entanglement of Matter and Meaning*. Durham, NC: Duke University Press, 2007.

Barbrook, Richard, and Andy Cameron. "The Californian Ideology." *Science as Culture* 6, no. 1 (1995): 44–72. https://doi.org/10.1080/09505439509526455.

Belasco, Warren J. *Appetite for Change: How the Counterculture Took on the Food Industry*. 2nd ed. Ithaca, NY: Cornell University Press, 2007.

Benson, Eric. "The War Against Beef." *Texas Monthly*, September 2020. https://www.texasmonthly.com/food/the-quest-for-better-beef/.

Bentley, Jeanine. "Trends in U.S. Per Capita Consumption of Dairy Products, 1970–2012." US Department of Agriculture Economic Research Service, June 2, 2014. https://www.ers.usda.gov/amber-waves/2014/june/trends-in-u-s-per-capita-consumption-of-dairy-products-1970-2012/.

Berlant, Lauren. "Big Man." *Social Text Online*, January 19, 2017. https://socialtextjournal.org/big-man/.

Berryman, Claire E., Harris R. Lieberman, Victor L. Fulgoni, and Stefan M. Pasiakos. "Protein Intake Trends and Conformity with the Dietary Reference Intakes in the United States: Analysis of the National Health and Nutrition Examination Survey, 2001–2014." *American Journal of Clinical Nutrition* 108, no. 2 (2018): 405–13. https://doi.org/10.1093/ajcn/nqy088.

Berson, Josh. *The Meat Question: Animals, Humans, and the Deep History of Food*. Cambridge, MA: MIT Press, 2019.

Berson, Josh. "Race and the Science of Starvation." *MIT Press Reader*, October 24, 2019. https://thereader.mitpress.mit.edu/race-and-the-science-of-starvation/.

Bertin, Oliver. "Ontario Program to Seek Uses from Wasted Whey." *Globe and Mail*, December 18, 1981.

Black, Maggie. *The Children and the Nations: The Story of* UNICEF. Sydney, Australia: UNICEF, 1986.

Blanchette, Alex. "Living Waste and the Labor of Toxic Health on American Factory Farms." *Medical Anthropology Quarterly* 33, no. 1 (2019): 80–100. https://doi.org/10.1111/maq.12491.

Blanchette, Alex. *Porkopolis: American Animality, Standardized Life, and the Factory Farm.* Durham, NC: Duke University Press, 2020.

Blaxter, Tamsin, and Tara Garnett. *Primed for Power: A Short Cultural History of Protein.* TABLE, University of Oxford, 2023. https://doi.org/10.56661/ba271ef5.

BMJ Best Practice. "Kwashiorkor." Last updated December 17, 2024. https://bestpractice .bmj.com/topics/en-us/1022.

Borre, Lisa. "Big Data Arrives." *National Geographic*, December 19, 2014. https://blog .nationalgeographic.org/2014/12/19/big-data-arrives-on-a-small- lake-in-vermont/.

Brantley, Cynthia. "Kikuyu-Maasai Nutrition and Colonial Science: The Orr and Gilks Study in Late 1920s Kenya Revisited." *International Journal of African Historical Studies* 30, no. 1 (1997): 49–86. https://doi.org/10.2307/221546.

Bratich, Jack. *On Microfascism: Gender, War, and Death.* Brooklyn, NY: Common Notions, 2022.

Bratich, Jack, and Sarah Banet-Weiser. "From Pick-Up Artists to Incels: Con Games, Networked Misogyny, and the Failure of Neoliberalism." *International Journal of Communication* 13 (September 2019): 5003–27.

Broad, Garrett. "Making Meat, Better: The Metaphors of Plant-Based and Cell-Based Meat Innovation." *Environmental Communication* 14, no. 7 (2020): 919–32. https://doi .org/10.1080/17524032.2020.1725085.

Brock, John, and Marcel Autret. *Kwashiorkor in Africa.* World Health Organization Monograph Series, no. 8. Geneva: World Health Organization, 1952. https://apps.who .int/iris/handle/10665/40717.

Brock, John, and J. Hansen. "Protein Requirement." *The Lancet* 304, no. 7882 (1974): 712–14.

Brock, William Hodson. *Justus von Liebig: The Chemical Gatekeeper.* Cambridge: Cambridge University Press, 1997.

Brown, Seth. "Alex Jones's Media Empire Is a Machine Built to Sell Snake-Oil Diet Supplements." *Intelligencer*, May 4, 2017. https://nymag.com/intelligencer/2017/05 /how-does-alex-jones-make-money.html.

Browne, Charles. "Justus von Liebig—Man and Teacher." In *Liebig and After Liebig: A Century of Progress in Agricultural Chemistry*, edited by Forest Ray Moulton. Washington, DC: American Association for the Advancement of Science, 1942.

Campbell, Nerissa, Stefanie De Jesus, and Harry Prapavessis. "Physical Fitness." In *Encyclopedia of Behavioral Medicine*, edited by Marc D. Gellman. New York: Springer, 2020. https://doi.org/10.1007/978-3-030-39903-0_1167.

Cannon, Geoffrey. "Nutrition: The New World Map." *Asia Pacific Journal of Clinical Nutrition* 11, no. 3 (2002): S480–97. https://doi.org/10.1046/j.1440-6047.11.supp3.4.x.

Cannon, Geoffrey. "The Rise and Fall of Dietetics and of Nutrition Science, 4000 B.C.E.–2000 C.E." *Public Health Nutrition* 8, no. 6a (2005): 701–5. https://doi.org/10 .1079/PHN2005766.

Carpenter, Kenneth J. "The History of Enthusiasm for Protein." *Journal of Nutrition* 116, no. 7 (1986): 1364–70. https://doi.org/10.1093/jn/116.7.1364.

Carpenter, Kenneth J. *Protein and Energy: A Study of Changing Ideas in Nutrition.* Cambridge: Cambridge University Press, 1994.

Carpenter, Kenneth J. "A Short History of Nutritional Science: Part 1 (1785–1885)." *Journal of Nutrition* 133, no. 3 (2003): 638–45. https://doi.org/10.1093/jn/133.3.638.

Cavallero, Lucía, and Verónica Gago. "Feminism, the Pandemic, and What Comes Next." Translated by Liz Mason-Deese. *Critical Times*, April 21, 2020. https://ctjournal.org/2020/04/21/feminism-the-pandemic-and-what-comes-next/.

Chakrabarty, Dipesh. *The Climate of History in a Planetary Age.* Chicago: University of Chicago Press, 2021.

Chen, Liang-Kung, Li-Kuo Liu, Jean Woo, Prasert Assantachai, Tung-Wai Auyeung, Kamaruzzaman Shahrul Bahyah, Ming-Yueh Chou, et al. "Sarcopenia in Asia: Consensus Report of the Asian Working Group for Sarcopenia." *Journal of Post Acute and Long-Term Care Medicine* 15, no. 2 (2014): 95–101. https://doi.org/10.1016/j.jamda.2013.11.025.

Chiles, Robert Magneson. "If They Come, We Will Build It: In Vitro Meat and the Discursive Struggle over Future Agrofood Expectations." *Agriculture and Human Values* 30, no. 4 (2013): 511–23. https://doi.org/10.1007/s10460-013-9427-9.

Chumlea, W. C., M. Cesari, W. J. Evans, L. Ferrucci, R. A. Fielding, M. Pahor, S. Studenski, B. Vellas, and the Task Force Members. "International Working Group on Sarcopenia." *Journal of Nutrition, Health and Aging* 15, no. 6 (2011): 450–55. https://doi.org/10.1007/s12603-011-0092-7.

Clark, Brian C., and Todd M. Manini. "What Is Dynapenia?" *Nutrition* 28, no. 5 (2012): 495–503. https://doi.org/10.1016/j.nut.2011.12.002.

Cole, Cheryl L., and Amy Hribar. "Celebrity Feminism: *Nike Style* Post-Fordism, Transcendence, and Consumer Power." *Sociology of Sport Journal* 12, no.4 (1995): 347–69. https://doi.org/10.1123/ssj.12.4.347.

Connell, Raewyn W. *Masculinities.* Berkeley: University of California Press, 1995.

Cook, Ian. "Follow the Thing: Papaya." *Antipode* 36, no. 4 (2004): 642–64. https://doi.org/10.1111/j.1467-8330.2004.00441.x.

Coole, Diana, and Samantha Frost, eds. *New Materialisms: Ontology, Agency, and Politics.* Durham, NC: Duke University Press, 2010. https://doi.org/10.1215/9780822392996.

Cooper, Melinda. "A Burden on Future Generations? How We Learned to Hate Deficits and Blame the Baby Boomers." *Sociological Review* 69, no. 4 (2021): 743–58. https://doi.org/10.1177/0038026121999211.

Cooper, Melinda. *Life as Surplus: Biotechnology and Capitalism in the Neoliberal Era.* Seattle: University of Washington Press, 2008.

Craddock, Sally. *Retired, Except on Demand: The Life of Cicely Williams.* Oxford, UK: Green College, 1983.

Crary, Jonathan. *Scorched Earth: Beyond the Digital Age to a Post-Capitalist World.* London: Verso, 2022.

Crawford, Robert. "Health as a Meaningful Social Practice." *Health: An Interdisciplinary Journal for the Social Study of Health, Illness and Medicine* 10, no. 4 (2006): 401–20. https://doi.org/10.1177/1363459306067310.

Crawford, Robert. "Healthism and the Medicalization of Everyday Life." *International Journal of Social Determinants of Health and Health Services* 10, no. 3 (1980): 365–88. https://doi.org/10.2190/3H2H-3XJN-3KAY-G9NY.

Creswell, Julie. "Beyond Meat Is Struggling, and the Plant-Based Meat Industry Worries." *New York Times*, November 21, 2022. https://www.nytimes.com/2022/11/21/business/beyond-meat-industry.html.

Crowfoot, James, and Julia M. Wondolleck. *Environmental Disputes: Community Involvement in Conflict Resolution.* Washington, DC: Island Press, 1990.

Cruz-Jentoft, Alfonso J., Jean-Pierre Baeyens, Jürgen M. Bauer, Yves Boirie, Tommy Cederholm, Francesco Landi, Finbarr C. Martin, et al. "Sarcopenia: European Consensus on Definition and Diagnosis: Report of the European Working Group on Sarcopenia in Older People." *Age Ageing* 39, no. 4 (2010): 412–23. https://doi.org/10.1093/ageing/afq034.

Cruz-Jentoft, Alfonso J., Gulistan Bahat, Jurgen Bauer, Yves Boirie, Olivier Bruyere, Tommy Cederholm, Cyrus Cooper, et al. "Sarcopenia: Revised European Consensus on Definition and Diagnosis." *Age Ageing* 48, no. 1 (2019): 16–31. https://doi.org/10.1093/ageing/afy169.

Davis, Garth, and Howard Jacobson. *Proteinaholic: How Our Obsession with Meat Is Killing Us and What We Can Do About It.* New York: HarperOne, 2016.

Davis, Janae, Alex A. Moulton, Levi Van Sant, and Brian Williams. "Anthropocene, Capitalocene, . . . Plantationocene? A Manifesto for Ecological Justice in an Age of Global Crises." *Geography Compass* 13, no. 5 (2019): e12438. http://doi.org/10.1111/gec3.12438.

Decker, Kimberly J. "How to Compete in the Crowded Protein Market." *Nutritional Outlook* 21, no. 9 (2019). https://www.nutritionaloutlook.com/view/how-compete-crowded-protein-market.

Deleuze, Gilles, and Félix Guattari. *A Thousand Plateaus: Capitalism and Schizophrenia.* Translated by Brian Massumi. Minneapolis: University of Minnesota Press, 1987.

de Rennie, Cecilia Molinari. "Advertising Capitalism: A Critical Analysis of Early 20th-Century LEMCO Trading Cards." *Bakhtiniana* 15, no. 4 (2020): 173–92. http://dx.doi.org/10.1590/2176-457347460.

Dewey, Caitlin. "America's Cheese Stockpile Just Hit an All-Time High." *Washington Post*, June 28, 2018. https://www.washingtonpost.com/news/wonk/wp/2018/06/28/americas-cheese-stockpile-just-hit-an-all-time-high/.

Douglas, Mary. *Purity and Danger: An Analysis of the Concepts of Pollution and Taboo.* Harmondsworth, UK: Penguin, 1970.

Drew, Liam. "Lifting the Burden of Old Age." *Nature (London)* 555, no. 7695 (2018): S15–S17. http://doi.org/10.1038/d41586-018-02479-z.

Dunn, Elizabeth G. "Why Grandma's Nutrition Drink Is So Hard to Disrupt." *New York Times*, last modified February 5, 2023. https://www.nytimes.com/2023/02/02/business/dealbook/nutrition-drinks-ensure-boost.html.

Dutkiewicz, Jan, and Gabriel Rosenberg. "Why Right Wingers Are So Afraid of Men Eating Vegetables." *New Republic*, April 17, 2023. https://newrepublic.com/article/171781/meat-culture-war-crickets.

Dworkin, Shari L., and Faye Linda Wachs. *Body Panic: Gender, Health, and the Selling of Fitness*. New York: New York University Press, 2009.

Erisman, Jan Willem, James N. Galloway, Nancy B. Dise, Mark A. Sutton, Albert Bleeker, Bruna Grizzetti, Allison M. Leach, and Wim de Vries. "Nitrogen: Too Much of a Vital Resource." World Wildlife Fund Network, April 2015. http://www.louisbolk.org/downloads/3005.pdf.

Evans, William J. "What Is Sarcopenia?" Special issue, *Journals of Gerontology: Series A* 50A (1995): 5–8. https://doi.org/10.1093/gerona/50a.special_issue.5.

Farrington, Caroline. "Take It with a Grain (or More) of Salt: Why Industry-Backed Dietary Guidelines Fail Americans and How to Fix Them." *University of Michigan Journal of Law Reform* 55, no. 2 (2022): 477–516. https://repository.law.umich.edu/mjlr/vol55/iss2/6.

Finlay, Mark R. "Early Marketing of the Theory of Nutrition: The Science and Culture of Liebig's Extract of Meat." In *Science and Culture of Nutrition, 1840–1940*, edited by Harmke Kamminga and Andrew Cunningham. Leiden: Brill, 1995. https://doi.org/10.1163/9789004418417_005.

Finlay, Mark R. "Quackery and Cookery: Justus von Liebig's Extract of Meat and the Theory of Nutrition in the Victorian Age." *Bulletin of the History of Medicine* 66, no. 3 (1992): 404–18.

Food and Agriculture Organization of the United Nations. *Lives in Peril: Protein and the Child*. Rome: FAO, 1970.

Foster, John Bellamy. *Marx's Ecology: Materialism and Nature*. New York: Monthly Review Press, 2000.

Foucault, Michel. *The Birth of Biopolitics: Lectures at the Collège de France, 1978–1979*. Translated by Graham Burchell. New York: Palgrave Macmillan, 2010.

Foucault, Michel. *The Birth of the Clinic: An Archaeology of Medical Perception*. Translated by A. M. Sheridan Smith. London: Routledge, 1973.

Foucault, Michel. *Discipline and Punish: The Birth of the Prison*. New York: Pantheon Books, 1977.

Fox News. "Jocko Willink Blasts COVID 'Experts' Lockdown Insistence." YouTube, posted April 1, 2021, 5 min., 19 sec., accessed May 30, 2023. https://www.youtube.com/watch?v=8ZVfVwvvkHQ&ab_channel=FoxNews.

Frost, Samantha. *Biocultural Creatures: Toward a New Theory of the Human*. Durham, NC: Duke University Press, 2016. https://doi.org/10.1215/9780822374350.

Gago, Verónica, and Marta Malo. "Introduction: The New Feminist Internationale." Translated by Liz Mason-Deese. Special issue, *South Atlantic Quarterly* 119, no. 3 (2020): 620–28.

Gambert, Iselin, and Tobias Linné. "From Rice Eaters to Soy Boys: Race, Gender, and Tropes of 'Plant Food Masculinity.'" *Animal Studies Journal* 7, no. 2 (2018): 129–79.

Gandhi, Leela. *Affective Communities: Anticolonial Thought, Fin-de-Siècle Radicalism, and the Politics of Friendship*. Durham, NC: Duke University Press, 2006.

Gard, Michael, and Jan Wright. *The Obesity Epidemic: Science, Morality, and Ideology*. London: Routledge, 2005.

Garrett, Reginald H., Charles M. Grisham, William G. Wilmore, and Imed E. Gallouzi. *Biochemistry*. Toronto: Cengage Learning, 2012.

Gibson-Graham, J. K. *A Postcapitalist Politics*. Minneapolis: University of Minnesota Press, 2006.

Gilks, J. L., and J. B. Orr. "The Nutritional Condition of the East African Native." *The Lancet (British Edition)* 209, no. 5402 (1927): 560–62. https://doi.org/10.1016/S0140-6736(00)75124-5.

Ging, Debbie. "Alphas, Betas, and Incels: Theorizing the Masculinities of the Manosphere." *Men and Masculinities* 22, no. 4 (2019): 638–57. https://doi.org/10.1177/1097184X17706401.

Glas, Eduard. "The Evolution of a Scientific Concept." *Journal for General Philosophy of Science* 30 (1999): 37–58. https://doi.org/10.1023/A:1008246507186.

Gleig, Ann, and Brenna Artinger. "#BuddhistCultureWars: BuddhaBros, Alt-Right Dharma, and Snowflake Sanghas." *Journal of Global Buddhism* 22, no. 1 (2021): 19–48. https://doi.org/10.5281/zenodo.4727561.

Goates, Scott, Kristy Du, Mary Beth Arensberg, Trudy Gaillard, Jack Guralnik, and Suzette L. Pereira. "Economic Impact of Hospitalizations in US Adults with Sarcopenia." *Journal of Frailty and Aging* 8, no. 2 (2019): 93–99. https://doi.org/10.14283/jfa.2019.10.

Gordon, Aubrey, and Michael Hobbes. "The Great Protein Fiasco." *Maintenance Phase* (podcast), August 31, 2021, 1 hour, 10 min. https://podcasts.apple.com/sg/podcast/the-great-protein-fiasco/id1535408667?i=1000533846093.

Grosz, Elizabeth. *Volatile Bodies: Toward a Corporeal Feminism*. Bloomington: Indiana University Press, 1994.

Gustafsson, Thomas, and Brun Ulfhake. "Sarcopenia: What Is the Origin of This Aging-Induced Disorder?" *Frontiers in Genetics* 12 (July 2021). https://doi.org/10.3389/fgene.2021.688526.

Guthman, Julie. "Binging and Purging: Agrofood Capitalism and the Body as Socioecological Fix." *Environment and Planning A: Economy and Space* 47, no. 12 (2015): 2522–36. https://doi.org/10.1068/a140005p.

Guthman, Julie. *The Problems with Solutions: Why Silicon Valley Can't Hack the Future of Food*. Berkeley: University of California Press, 2024.

Guthman, Julie, and Charlotte Biltekoff. "Agri-Food Tech's Building Block: Narrating Protein, Agnostic of Source, in the Face of Crisis." *BioSocieties* 18, no. 3 (2023): 656–78.

Guthman, Julie, and Charlotte Biltekoff. "Magical Disruption? Alternative Protein and the Promise of De-Materialization." *Environment and Planning E: Nature and Space* 4, no. 4 (2021): 1583–1600. https://doi.org/10.1177/2514848620963125.

Guthman, Julie, and Madeline Fairbairn. "Speculating on Collapse: Unrealized Socioecological Fixes of Agri-Food Tech." *Environment and Planning A: Economy and Space* 56, no. 8 (2024): 2055–69. https://doi.org/10.1177/0308518X241265283.

Haase, Christoffer Bjerre, John Brandt Brodersen, and Jacob Bülow. "Sarcopenia: Early Prevention or Overdiagnosis?" *BMJ (Online)* 376 (2022): e052592. https://doi.org/10.1136/bmj-2019-052592.

Hacking, Ian. *Historical Ontology*. Cambridge, MA: Harvard University Press, 2004.

Haigh, Laxmi. "Protein Boost for Boost: Nestlé Ups the Content in Flagship 'Adult Nutrition' Product." *Nutrition Insight*, August 14, 2018. https://www.nutritioninsight

.com/news/protein-boost-for-boost-nestle-ups-the-content-in-flagship-adult
-nutrition-product.html.

Haraway, Donna J. *When Species Meet*. Minneapolis: University of Minnesota Press, 2008.

Hargreaves, Jennifer, and Patricia Vertinsky, eds. *Physical Culture, Power and the Body*. London: Routledge, 2006.

Harnish, Amelia. "How to Build a Better Bro." *Refinery 29*, last modified September 23, 2019. https://www.refinery29.com/en-us/2019/09/8448957/aubrey-marcus-podcast -philosopher-spirtuality-psychedelics.

Haushofer, Lisa. "Between Food and Medicine: Artificial Digestion, Sickness, and the Case of Benger's Food." *Journal of the History of Medicine and Allied Sciences* 73, no. 2 (2018): 168–87. https://doi.org/10.1093/jhmas/jry009.

Havstad, Joyce. "Protein Tokens, Types, and Taxa." In *Natural Kinds and Classification in Scientific Practice*, edited by Catherine Kendig. Abingdon, UK: Routledge, 2016.

Hayes, Maria. "Measuring Protein Content in Food: An Overview of Methods." *Foods* 9, no. 10 (2020): 1340. https://doi.org/10.3390/foods9101340.

Heffernan, Conor. "Physical Culture Supplements in Early Twentieth Century Britain." *Food Studies: An Interdisciplinary Journal* 12, no. 1 (2022): 53–68. https://doi.org/10 .18848/2160-1933/CGP/v12i01/53-68.

Heid, Markham. "The Great Protein Debate Heats Up." *Medium*, January 6, 2020. https://heated.medium.com/the-great-protein-debate-rages-on-f87ca89d556.

Heldke, Lisa. "It's Chomping All the Way Down: Toward an Ontology of the Human Individual." *The Monist* 101, no. 3 (2018): 247–60.

Herzog, Walter. "Skeletal Muscle Mechanics: Questions, Problems and Possible Solutions." *Journal of NeuroEngineering and Rehabilitation* 14, no. 98 (2017). https://doi .org/10.1186/s12984-017-0310-6.

Hesman-Saey, Tina. "Has AlphaFold Actually Solved Biology's Protein-Folding Problem?" *Sciencenews.org*, September 22, 2022. https://www.sciencenews.org/article /alphafold-ai-protein-structure-folding-prediction.

Homayun, Omaid. "How Aubrey Marcus Built Onnit into a $28-Million Company in 5 Years." *Forbes*, April 6, 2016. https://www.forbes.com/sites/omaidhomayun/2016/04 /06/how-aubrey-marcus-built-onnit-into-a-28-million-company-in-5-years/.

Hosie, Rachel. "Soy Boy: What Does This Online Insult Mean." *Independent*, September 30, 2020. https://www.independent.co.uk/life-style/soy-boy-insult-what-is -definition-far-right-men-masculinity-women-a8027816.html.

Howard, Philip H., Francesco Ajena, Marina Yamaoka, and Amber Clarke. "'Protein' Industry Convergence and Its Implications for Resilient and Equitable Food Systems." *Frontiers in Sustainable Food Systems* 5, no. 684181 (2021). https://doi.org/10 .3389/fsufs.2021.684181.

Howell, Jeremy, and Alan Ingham. "From Social Problem to Personal Issue: The Language of Lifestyle." *Cultural Studies* 15, no. 2 (2001): 326–51. https://doi.org/10 .1080/09502380152390535.

Howes, Laura. "Many of Our Proteins Remain Hidden in the Dark Proteome." *Chemical and Engineering News* 100, no. 3 (2022). https://cen.acs.org/biological-chemistry /proteomics/proteins-remain-hidden-dark-proteome/100/i3.

Immen, Wallace. "Human-Life Cow's Milk Is Sought." *Globe and Mail*, November 20, 1984.

Individuation Portal. "Aubrey Marcus and Joe Rogan—'Ruthless Self Love.'" YouTube, December 15, 2020, 2 min., 21 sec. https://youtu.be/_AfOKLRk7mU.

"Infographic: Canada's Seniors Population Outlook: Uncharted Territory." *Canadian Institute for Health Information*, 2017. https://www.cihi.ca/en/infographic-canadas -seniors-population-outlook-uncharted-territory.

Ingham, Alan G. "From Public Issue to Personal Trouble: Well-Being and the Fiscal Crisis of the State." *Sociology of Sport Journal* 2, no. 1 (1985): 43–55. https://doi.org/10 .1123/ssj.2.1.43.

Institute of Medicine, Panel on Macronutrients. *Dietary Reference Intakes for Energy, Carbohydrate, Fiber, Fat, Fatty Acids, Cholesterol, Protein and Amino Acids*. Washington, DC: National Academic Press, 2002. https://doi.org/10.17226/10490.

Jacobsen, Rowan. "Artificial Proteins Never Seen in the Natural World Are Becoming New COVID Vaccines and Medicines." *Scientific American*, July 1, 2021. https://www .scientificamerican.com/article/artificial-proteins-never-seen-in-the-natural-world -are-becoming-new-covid-vaccines-and-medicines/.

Jane, Emma A. *Misogyny Online: A Short (and Brutish) History*. London: Sage, 2017.

Janssen, Ian, Donald S. Shepard, Peter T. Katzmarzyk, and Ronenn Roubenoff. "The Healthcare Costs of Sarcopenia in the United States." *Journal of the American Geriatrics Society* 52, no. 1 (2004): 80–85. https://doi.org/10.1111/j.1532-5415.2004 .52014.x.

Jeffords, Susan. *Hard Bodies: Hollywood Masculinity in the Reagan Era*. New Brunswick, NJ: Rutgers University Press, 1994.

Jelen, Pavel. "Whey Processing—Utilization and Products." In *Encyclopedia of Dairy Sciences*, edited by H. Roginski, J. Fuquay, and P. Fox. London: Academic Press, 2003.

Jelliffe, D. B. "Commerciogenic Malnutrition?" *Nutrition Reviews* 30, no. 9 (1972): 199–205. https://doi.org/10.1111/j.1753-4887.1972.tb04042.x.

Jhally, Sut. *The Codes of Advertising: Fetishism and the Political Economy of Meaning in the Consumer Society*. New York: St. Martin's, 1987.

Jingnan, Huo. "How a Conspiracy Theory About Eating Bugs Made Its Way to International Politics." National Public Radio, March 31, 2023. https://www.npr.org /2023/03/31/1167550482/how-a-conspiracy-theory-about-eating-bugs-made-its-way-to -international-politics.

Jönsson, Erik. "Benevolent Technotopias and Hitherto Unimaginable Meats: Tracing the Promises of In Vitro Meat." *Social Studies of Science* 46, no. 5 (2016): 725–48. https://doi.org/10.1177/0306312716658561.

Jönsson, Erik, Tobias Linné, and Ally McCrow-Young. "Many Meats and Many Milks? The Ontological Politics of a Proposed Post-Animal Revolution." *Science as Culture* 28, no. 1 (2019): 70–97. https://doi.org/10.1080/09505431.2018.1544232.

Karera, Axelle. "Blackness and the Pitfalls of Anthropocene Ethics." *Critical Philosophy of Race* 7, no. 1 (2019): 32–56. https://doi.org/10.5325/critphilrace.7.1.0032.

Kelly, Megan, Alex DiBranco, and Julia R. DeCook. "Misogynist Incels and Male Supremacism: Overview and Recommendations for Addressing the Threat of Male

Supremacist Violence." *New America*, February 18, 2021. https://www.newamerica.org
/political-reform/reports/misogynist-incels-and-male-supremacism/.

Kenny, Katherine E. "The Biopolitics of Global Health: Life and Death in
Neoliberal Time." *Journal of Sociology* 51, no. 1 (2015): 9–27. https://doi.org/10.1177
/1440783314562313.

Kimura, Aya Hirata. *Hidden Hunger: Gender and the Politics of Smarter Foods*. Ithaca, NY:
Cornell University Press, 2013.

King, Samantha, and Gavin Weedon. "Embodiment Is Ecological: The Metabolic Lives
of Whey Protein Powder." *Body and Society* 26, no. 1 (2020): 82–106. https://doi.org/10
.1177/1357034X19878775.

King, Samantha, and Gavin Weedon. "The Nature of the Body in Sport and Physical
Culture: From Bodies and Environments to Ecological Embodiment." *Sociology of
Sport Journal* 38, no. 2 (2021): 131–39. https://doi.org/10.1123/ssj.2020-0038.

Kirksey, S. Eben, and Stefan Helmreich. "The Emergence of Multispecies Ethnography."
Cultural Anthropology 25, no. 4 (2010): 545–76. https://doi.org/10.1111/j.1548-1360.2010
.01069.x.

Knight, Michael. "Whey Disposal Troubles Vermonters." *New York Times*, June 20, 1979.
https://www.nytimes.com/1979/06/25/archives/whey-disposal-troubles-vermonters
-surprised-at-surprise.html.

Konotey-Ahulu, Felix D. "There Is Nothing Mysterious About Kwashiorkor" (Rapid
Response to "Antioxidants for Children with Kwashiorkor"). *British Medical Journal*
330, no. 1095 (2005). https://www.bmj.com/rapid-response/2011/10/30/there-nothing
-mysterious-about-kwashiorkor.

Konotey-Ahulu, Felix D., and Donald S. McLaren. "Issues in Kwashiorkor." *The Lancet*
343, no. 8896 (1994): 548–49. https://doi.org/10.1016/S0140-6736(94)91504-0.

Kuhn, Thomas, S. *The Structure of Scientific Revolutions: 50th Anniversary Edition*.
Chicago: University of Chicago Press, 2012.

Kusz, Kyle, and Matthew R. Hodler. "'Saturdays Are for the Boys': Barstool Sports and
the Cultural Politics of White Fratriarchy in Contemporary America." *Sociology of
Sport Journal* 40, no. 1 (2022): 1–12. https://doi.org/10.1123/ssj.2022-0075.

Landecker, Hannah. "A Metabolic History of Manufacturing Waste: Food Commodities
and Their Outsides." *Food, Culture and Society* 22, no. 5 (2019): 530–47. https://doi.org
/10.1080/15528014.2019.1638110.

Lappe, Frances Moore. *Diet for a Small Planet*. New York: Ballantine Books, 1971.

Latour, Bruno. *Reassembling the Social: An Introduction to Actor-Network-Theory*. Oxford:
Oxford University Press, 2005.

Latour, Bruno. *We Have Never Been Modern*. Cambridge, MA: Harvard University Press,
1993.

Ledin, Erik. "The Protein Interview: An interview with Dr. Stuart Phillips." *Lean Bodies
Consulting*, April 17, 2014. https://leanbodiesconsulting.com/articles/the-protein
-interview-an-interview-with-dr-stuart-phillips/.

Leong, Diana. "The Mattering of Black Lives: Octavia Butler's Hyperempathy and the
Promise of the New Materialisms." *Catalyst: Feminism, Theory, Technoscience* 2, no. 2
(2016): 1–35. https://doi.org/10.28968/cftt.v2i2.28799.

Lewowicz, Lucia. LEMCO: *Un coloso de la industria cárnica en Fray Bentos, Uruguay*. Montevideo, Uruguay: INAC, 2016.

Liboiron, Max. "Redefining Pollution and Action: The Matter of Plastics." *Journal of Material Culture* 21, no. 1 (2016): 87–110. https://doi.org/10.1177/1359183515622966.

Liebig, Justus von. "Animal Chemistry; or Organic Chemistry in Its Applications to Physiology and Pathology." *Medico-Chirurgical Review* 37, no. 74 (1842): 337–71.

Liebig, Justus von. "The Nutritive Value of Different Sorts of Food." *The Lancet* 93, no. 2366 (1869): 4–5.

Liebig, Justus von. *Researches on the Chemistry of Food*. London: Taylor and Walton, 1847.

Lonkila, Annika, and Minna Kaljonen. "Promises of Meat and Milk Alternatives: An Integrative Literature Review on Emergent Research Themes." *Agriculture and Human Values* 38 (2021): 625–39. https://doi.org/10.1007/s10460-020-10184-9.

Lougheed, Scott. "An Actor-Network Theory Examination of Cheese and Whey Production in Ontario." Master's thesis, Queen's University, Kingston, Ontario, 2013.

Luiz, Stephanie. "My Beef with Dairy: How the US Government Is Bailing Out a Dying Industry." *Northeastern University Political Review*, May 16, 2020. https://nupoliticalreview.org/2020/05/16/my-beef-with-dairy-how-the-us-government-is-bailing-out-a-dying-industry/.

Lynch, John, and Raymond Pierrehumbert. "Climate Impacts of Cultured Meat and Beef Cattle." *Frontiers in Sustainable Food Systems*, February 18, 2019. https://www.frontiersin.org/journals/sustainable-food-systems/articles/10.3389/fsufs.2019.00005/full.

MacDonald, Katie. "Producing Protein." *Canadian Food Studies* 11, no. 1 (2024): 30–46. https://doi.org/10.15353/cfs-rcea.v11i1.635.

Marcell, Taylor J. "Sarcopenia: Causes, Consequences, and Preventions." *Journals of Gerontology: Series A* 58, no. 10 (2003): M911–M916. https://doi.org/10.1093/gerona/58.10.M911.

Marchesi, Greta. "Justus von Liebig Makes the World: Soil Properties and Social Change in the Nineteenth Century." *Environmental Humanities* 12, no. 1 (2020): 205–26. https://doi.org/10.1215/22011919-8142308.

Marcus, Aubrey. *Own the Day, Own Your Life: Optimized Practices for Waking, Working, Learning, Eating, Training, Playing, Sleeping and Sex*. New York: Harper Wave, 2018.

Marcus, Aubrey. "The Sacred Masculine Archetype." *Aubrey Marcus Podcast*, May 5, 2021. https://www.aubreymarcus.com/blogs/aubrey-marcus-podcast/the-sacred-masculine-archetype-with-erick-godsey-kyle-kingsbury-amp-307.

Martschukat, Jürgen. *The Age of Fitness: How the Body Came to Symbolize Success and Achievement*. Translated by Alex Skinner. Cambridge, UK: Polity, 2021.

Marx, Karl. *Capital: A Critique of Political Economy*. Vol. 1. Translated by Ben Fowkes. London: Penguin Classics, 1992.

Marx, Karl. *Capital: A Critique of Political Economy*. Vol. 3. Translated by Ernest Mandel. London: Penguin Classics, 1993

Mbembe, Achille. "Futures of Life and Futures of Reason." *Public Culture* 33, no. 1 (2021): 11–33. https://doi.org/10.1215/08992363-8742136.

McGuire-Adams, Tricia. *Indigenous Feminist Gikendaasowin (Knowledge): Decolonization through Physical Activity*. Cham, Switzerland: Palgrave MacMillan, 2021.

McLaren, Donald S. "A Fresh Look at Protein-Calorie Malnutrition." *The Lancet* 288, no. 7461 (1966): 485–88.

McLaren, Donald S. "The Great Protein Fiasco." *The Lancet* 304, no. 7872 (1974): 93–96.

McLaren, Donald S. "The Great Protein Fiasco Revisited." *Nutrition* 16, no. 6 (2000): 464–65. https://doi.org/10.1016/S0899-9007(00)00234-3.

McMichael, Philip. "A Food Regime Genealogy." *Journal of Peasant Studies* 36, no. 1 (2009): 139–69. https://doi.org/10.1080/03066150902820354.

Mendel, Lafayette B. *Nutrition: The Chemistry of Life*. New Haven, CT: Yale University Press, 1923.

Menzies, Heather. *By the Labor of Their Hands: The Story of Ontario Cheddar Cheese*. Kingston, ON: Quarry Press, 1994.

Messner, Michael A. *Politics of Masculinities: Men in Movements*. Thousand Oaks, CA: Sage, 1997.

Metzl, Jonathan M., and Anna Kirkland. *Against Health: How Health Became the New Morality*. New York: New York University Press, 2010.

Millington, Brad. *Fitness, Technology and Society: Amusing Ourselves to Life*. Abingdon, UK: Routledge, 2018.

Mol, Annemarie. *The Body Multiple: Ontology in Medical Practice*. Durham, NC: Duke University Press, 2002.

Mol, Annemarie. "Layers or Versions? Human Bodies and the Love of Bitterness." In *The Routledge Handbook of the Body*, edited by Brian Turner. Oxford, UK: Routledge, 2012.

Moore, Jason W. *Capitalism in the Web of Life: Ecology and the Accumulation of Capital*. London: Verso, 2015.

Moran, Susan K. "Wastewater Is Key to Reducing Nitrogen Pollution." *Scientific American*, June 2, 2016. https://www.scientificamerican.com/article/wastewater-is-key-to-reducing-nitrogen-pollution/.

Moss, Michael. "While Warning About Fat, US Pushes Cheese Sales." *New York Times*, November 7, 2010. https://www.nytimes.com/2010/11/07/us/07fat.html.

Moulton, Forest Ray, ed. *Liebig and After Liebig: A Century of Progress in Agricultural Chemistry*. Washington, DC: American Association for the Advancement of Science, 1942.

Mulder, Arnold. "The Quest for Sustainable Nitrogen Removal Technologies." *Water Science and Technology* 48, no. 1 (2003): 67–75.

Munday, Pat. "Politics by Other Means: Justus von Liebig and the German Translation of John Stuart Mill's *Logic*." *British Journal for the History of Science* 31, no. 4 (1998): 403–18. https://doi.org/10.1017/S0007087498003379.

Murphy, M. *The Economization of Life*. Durham, NC: Duke University Press, 2017.

Myers, Natasha. *Rendering Life Molecular: Models, Modelers, and Excitable Matter*. Durham, NC: Duke University Press, 2015.

Mylan, Josephine, John Andrews, and Damian Maye. "The Big Business of Sustainable Food Production and Consumption: Exploring the Transition to Alternative Proteins." *Proceedings of the National Academic of Sciences of the United States of America* 120 (47) (2023): e2207782120. https://doi.org/10.1073/pnas.2207782120.

Nishimura, Yusuke, Grith Højfeldt, Leigh Breen, Inge Tetens, and Lars Holm. "Dietary Protein Requirements and Recommendations for Healthy Older Adults: A Critical Narrative Review of the Scientific Evidence." *Nutrition Research Reviews* 36, no. 1 (2023): 69–85. https://doi.org/10.1017/S0954422421000329.

Nkhoma, Bryson. "The 'Malnutrition Syndrome': African Diets, Nutrition Science, and Colonial Research in Southern Africa." *African Historical Review* 54, no. 2 (2023): 1–12. https://doi.org/10.1080/17532523.2022.2153439.

Nkhoma, Bryson G. "'We Are What We Eat': Nutrition, African Diets and the State in Colonial Malawi, 1920s–1960." *Journal of Southern African Studies* 46, no. 6 (2020): 1219–35. https://doi.org/10.1080/03057070.2020.1830548.

Nott, John. "'How Little Progress'? A Political Economy of Postcolonial Nutrition." *Population and Development Review* 44, no. 4 (2018): 771–91. https://doi.org/10.1111/padr.12198.

Nott, John. "'No One May Starve in the British Empire': Kwashiorkor, Protein and the Politics of Nutrition Between Britain and Africa." *Social History of Medicine* 34, no. 2 (2021): 553–76. https://doi.org/10.1093/shm/hkz107.

Onwulata, Charles I., and Peter J. Huth, eds. *Whey Processing, Functionality and Health Benefits*. Ames, IA: Wiley-Blackwell, 2008.

Oppenheim, Maya. "Figures That Lay Bare the Shocking Scale of Toxic Influencer Andrew Tate's Reach Among Young Men." *Independent*, February 16, 2023. https://www.independent.co.uk/news/uk/home-news/andrew-tate-influence-young-men-misogyny-b2283595.html.

Orr, J. B., and J. L. Gilks. *Studies of Nutrition: The Physique and Health of Two African Tribes*. London: H. M. Stationery Office, 1931.

Osaka, Shannon. "The Big Problem with Plant-Based Meat: The Meat Part." *Washington Post*, January 19, 2023. https://www.washingtonpost.com/climate-solutions/2023/01/19/plant-based-meat-failing/.

Overend, Alissa. *Shifting Food Facts: Dietary Discourse in a Post-Truth Culture*. London: Routledge, 2021.

Page, Arnaud. "'The Greatest Victory Which the Chemist Has Won in the Fight (. . .) Against Nature': Nitrogenous Fertilizers in Great Britain and the British Empire, 1910s–1950s." *History of Science* 54, no. 4 (2016): 383–98. https://doi.org/10.1177/0073275316681801.

Patchin, Paige Marie. "For the Sake of the Child: The Economisation of Reproduction in the Zika Public Health Emergency." *Transactions of the Institute of British Geographers* 46, no. 1 (2021): 2–14. https://doi.org/10.1111/tran.12384.

Paxson, Heather. *The Life of Cheese: Crafting Food and Value in America*. Berkeley: University of California Press, 2012.

Pearson, Jordan. "Inside the World of 'Bitcoin Carnivores.'" *Vice*, September 29, 2017. https://www.vice.com/en/article/ne74nw/inside-the-world-of-the-bitcoin-carnivores.

Pencharz, P., F. Jahoor, A. Kurpad, K. F. Michaelsen, C. Slater, D. Tome, and R. Weisell. "Current Issues in Determining Dietary Protein and Amino-Acid Requirements." *European Journal of Clinical Nutrition* 68, no. 3 (2014): 285–86. https://doi.org/10.1038/ejcn.2013.297.

Peterson, Jordan B. *Beyond Order: 12 More Rules for Life*. New York: Penguin, 2021.

Pollan, Michael. *In Defense of Food: An Eater's Manifesto*. New York: Penguin Press, 2008.

Poulson, Brittany, Heather M. Trevino, and Rita Sather. "Leucine." University of Rochester Medical Center Health Encyclopedia, accessed October 4, 2023. https://www.urmc.rochester.edu/encyclopedia/content.aspx?contenttypeid=19&contentid=Leucine.

Proceedings of the Physiological Society. "Protein Nomenclature." *Proceedings of the Physiological Society*, January 26, 1907, xvii–xx.

Pulidindi, Kiran, and Kunal Ahuja. "Protein Powder Market Size by Source." *Global Market Insights*, 2022. https://www.gminsights.com/industry-analysis/protein-powder-market.

Purdy, Chase. "Functional Foods Are Boring. Someone Tell Silicon Valley." *Quartz*, September 1, 2018. https://qz.com/quartzy/1375904/functional-foods-are-boring -someone-tell-silicon-valley.

Reisenwitz, Cathy. "Do You Even Trip, Bro? On Psychedelic Masculinity." *Psychedelic*, August 25, 2022. https://psychedelicspotlight.com/do-you-even-trip-bro-on -psychedelic-masculinity/.

Remski, Matthew. "Bro Science Manifesto: Hot Male Conspirituality Seeks Credentials." *Medium*, July 7, 2021. https://matthewremski.medium.com/bro-science-manifesto -b6f6ec7d5481#:~:text=By%20contrast%2C%20the%20banal%20health,is%20the%20 pathway%20to%20virtue.

Rivers, J. P. W. "The Profession of Nutrition—An Historical Perspective." *Proceedings of the Nutrition Society* 38, no. 2 (1979): 225–31. https://doi.org/10.1079/PNS19790035.

Rose, Nikolas. "Molecular Biopolitics, Somatic Ethics and the Spirit of Biocapital." *Social Theory and Health* 5 (2007): 3–29. https://doi.org/10.1057/palgrave.sth.8700084.

Rose, Nikolas. *The Politics of Life Itself: Biomedicine, Power, and Subjectivity in the Twenty-First Century*. Princeton, NJ: Princeton University Press, 2007.

Rosenberg, Irwin H. "Sarcopenia: Origins and Clinical Relevance." *Journal of Nutrition* 127, no. 5 (1997): 990S–91S. https://doi.org/10.1093/jn/127.5.990S.

Rosenberg, Irwin H. "Sarcopenia: Origins and Clinical Relevance." *Clinics in Geriatric Medicine* 27, no. 3 (2011): 337–39. https://doi.org/10.1016/j.cger.2011.03.003.

Roy, Parama. *Alimentary Tracts: Appetites, Aversions, and the Postcolonial*. Durham, NC: Duke University Press, 2010.

Sai, Frederick. "Nutrition in Ghana." In *Nutrition and National Policy*, edited by Beverly Winikoff. Cambridge, MA: MIT Press, 1978.

Santo, Raychel E., Brent F. Kim, Sarah E. Goldman, Jan Dutkiewicz, Erin M. B. Biehl, W. Bloem, Roni A. Neff, and Keeve E. Nachman. "Considering Plant-Based Meat Substitutes and Cell-Based Meats: A Public Health and Food Systems Perspective." *Frontiers in Sustainable Food Systems* 4 (August 2020). https://doi.org/10.3389/fsufs.2020.00134.

Sarmiento, Eric. "Umwelt, Food, and the Limits of Control." *Emotion, Space and Society* 14, no. 1 (2015): 74–83. https://doi.org/10.1016/j.emospa.2013.08.008.

Sathyamala, C. "Nutritionalizing Food: A Framework for Capital Accumulation." *Development and Change* 47, no. 4 (2016): 818–39. https://doi.org/10.1111/dech.12250.

Scanlon, John. *On Garbage*. Chicago: University of Chicago Press, 2005.

Schöley, Jonas, Jose Manuel Aburto, Ilya Kashnitsky, Maxi S. Kniffka, Luyin Zhang, Hannaliis Jaadla, Jennifer B. Dowd, and Ridhi Kashyap. "Life Expectancy Changes

Since COVID-19." *Nature Human Behaviour* 6 (2022): 1649–59. https://doi.org/10.1038/s41562-022-01450-3.

Schrecker, Ted. "Neoliberalism and Health: The Linkages and the Dangers." *Sociology Compass* 10, no. 10 (2016): 952–71. https://doi.org/10.1111/soc4.12408.

Scott, John. "Domtar Paper Mill Termed Major Trent Polluter." *Globe and Mail*, February 3, 1971.

Scott-Reid, Jessica. "Opinion: 'Ultraprocessed' Plant-Based Meat Isn't as Bad for You as the Meat Industry Wants You to Believe." *Toronto Star*, last updated April 28, 2025. https://www.thestar.com/opinion/contributors/ultraprocessed-plant-based-meat-isnt-as-bad-for-you-as-the-meat-industry-wants-you/article_7cd5cb1e-3944-11ef-98a3-630c7eb74f1d.html.

Scott-Smith, Tom. "The Fetishism of Humanitarian Objects and the Management of Malnutrition in Emergencies. *Third World Quarterly* 34, no.5 (2013): 913–28. https://doi.org/10.1080/01436597.2013.800749.

Scott-Smith, Tom. *On an Empty Stomach*. Ithaca, NY: Cornell University Press, 2020.

Scrinis, Gyorgy. *Nutritionism: The Science and Politics of Dietary Advice*. New York: Columbia University Press, 2013.

Semba, Richard D. "The Rise and Fall of Protein Malnutrition in Global Health." *Annals of Nutrition and Metabolism* 69, no. 2 (2016): 79–88. https://doi.org/10.1159/000449175.

"Senior Nutrition: Personalization, Protein and Genderfication as Clear Drivers." *Nutrition Insight*, July 5, 2018. https://www.nutritioninsight.com/news/senior-nutrition-personalization-protein-and-genderfication-as-clear-drivers.html.

Service, Robert F. "'The Game Has Changed': AI Triumphs at Solving Protein Structures." *Science*, November 30, 2020. https://www.science.org/content/article/game-has-changed-ai-triumphs-solving-protein-structures.

Sexton, Alexandra E. "Eating for the Post-Anthropocene: Alternative Proteins and the Biopolitics of Edibility." *Transactions of the Institute of British Geographers* 43, no. 4 (2018): 586–600. https://doi.org/10.1111/tran.12253.

Sexton, Alexandra E. "Food as Software: Place, Protein, and Feeding the World Silicon Valley–Style." *Economic Geography* 96, no. 5 (2020): 449–69. https://doi.org/10.1080/00130095.2020.1834382.

Sexton, Alexandra E., Tara Garnett, and Jamie Lorimer. "Framing the Future of Food: The Contested Promises of Alternative Proteins." *Environment and Planning E: Nature and Space* 2, no. 1 (2019): 47–72. https://doi.org/10.1177/2514848619827009.

Seymour, Richard. *Disaster Nationalism: The Downfall of Liberal Civilization*. London: Verso, 2024.

Seymour, Richard. *The Twittering Machine*. London: Indigo Press, 2019.

Shalal, Andrea. "Aging Population to Hit U.S. Economy like a 'Ton of Bricks'—U.S. Commerce Secretary." Reuters, last modified July 12, 2021. https://www.reuters.com/article/us-usa-economy-raimondo-idTRNIKBN2EI29F.

Shotwell, Alexis. *Against Purity: Living Ethically in Compromised Times*. Minneapolis: University of Minnesota Press, 2016.

Smith, Robyn. "The Emergence of Vitamins as Bio-Political Objects During World War I." *Studies in History and Philosophy of Science Part C: Studies in History and*

Philosophy of Biological and Biomedical Sciences 40, no. 3 (2009): 179–89. https://doi.org
/10.1016/j.shpsc.2009.06.006.

Smithers, Geoffrey W. "Whey and Whey Proteins—From 'Gutter-to-Gold.'"
International Dairy Journal 18, no. 7 (2008): 695–704. https://doi.org/10.1016/j.idairyj
.2008.03.008.

Spencer, Dale C. "'Eating Clean' for a Violent Body: Mixed Martial Arts, Diet and
Masculinities." *Women's Studies International Forum* 44 (2014): 247–54. https://doi.org
/10.1016/j.wsif.2013.05.018.

Stanton, J. "Listening to the Ga: Cicely William's Discovery of Kwashiorkor on the Gold
Coast." In *Women and Modern Medicine*, edited by Laurence Conrad and Anne Hardy.
Clio Medica no. 61. Leiden: Brill, 2001. https://brill.com/display/title/28321.

Steffen, Will, Paul J. Crutzen, and John R. McNeill. "The Anthropocene: Are
Humans Now Overwhelming the Great Forces of Nature?" *AMBIO: A Journal of
the Human Environment* 36, no. 8 (2007): 614–21. https://doi.org/10.1579/0044
-7447(2007)36[614:TAAHNO]2.0.CO;2.

Stephens, Neil, Lucy Di Silvio, Illtud Dunsford, Marianne Ellis, Abigail Glencross, and
Alexandra Sexton. "Bringing Cultured Meat to Market: Technical, Socio-Political, and
Regulatory Challenges in Cellular Agriculture." *Trends in Food Science and Technology*
78 (2018): 155–66. https://doi.org/10.1016/j.tifs.2018.04.010.

Stephens, Neil, Emma King, and Catherine Lyall. "Blood, Meat, and Upscaling Tissue
Engineering: Promises, Anticipated Markets, and Performativity in the Biomedical
and Agri-Food Sectors." *BioSocieties*, 13, no. 2 (2018): 368–88. https://doi.org/10.1057
/s41292-017-0072-1.

Sugiura, Lisa. *The Incel Rebellion: The Rise of the Manosphere and the Virtual War Against
Women.* Bingley, UK: Emerald Publishing, 2021. https://library.oapen.org/handle/20
.500.12657/51536.

Sunder Rajan, Kaushik. *Biocapital: The Constitution of Postgenomic Life.* Durham, NC:
Duke University Press, 2006.

Sunseri, Thaddeus. "International Beef Packing in the Age of Empire: LEMCO in South
West Africa, 1906–c. 1940." *South African Historical Journal* 73, no. 3 (2021): 573–600.
https://doi.org/10.1080/02582473.2021.1965200.

TallBear, Kim. "An Indigenous Reflection on Working Beyond the Human/Not Human."
GLQ: A Journal of Lesbian and Gay Studies 21, no. 2 (2015): 230–35.

Tanford, Charles, and Jacqueline Reynolds. *Nature's Robots: A History of Proteins.* New
York: Oxford University Press, 2003.

Temple, James. "Bill Gates: Rich Nations Should Shift Entirely to Synthetic Beef." *MIT
Technology Review*, February 14, 2021. https://www.technologyreview.com/2021/02/14
/1018296/bill-gates-climate-change-beef-trees-microsoft/.

Toffoletti, Kim, and Holly Thorpe. "Bodies, Gender, and Digital Affect in Fitspiration
Media." *Feminist Media Studies* 21, no. 5 (2021): 822–39. https://doi.org/10.1080
/14680777.2020.1713841.

Trowell, H. C. "Malignant Malnutrition." *Transactions of the Royal Society of Tropical
Medicine and Hygiene* 42, no. 5 (1949): 417–42. https://doi.org/10.1016/0035
-9203(49)90049-8.

Trowell, H. C., and J. N. P. Davis. "Kwashiorkor—I. Nutritional Background, History, and Distribution." *British Medical Journal* 2, no. 2788 (1952): 796–98. https://doi.org/10.1136/bmj.2.4788.796.

Truog, Robert D., Nancy Berlinger, Rachel L. Zacharias, and Mildred Z. Solomon. "Brain Death at Fifty: Exploring Consensus, Controversy, and Contexts." *Hastings Center Report* 48, no. 6 (2018): S2–S5. https://doi.org/10.1002/hast.942.

Tunick, Michael H. "Whey Protein Production and Utilization: A Brief History." In *Whey Processing, Functionality and Health Benefits*, edited by Charles I. Onwulata and Peter J. Huth. Ames, IA: Blackwell Publishing and the Institute of Food Technologists, 2008.

Turner, Fred. *From Counterculture to Cyberculture: Stewart Brand, the Whole Earth Network and the Rise of Digital Utopianism*. Chicago: University of Chicago Press, 2006.

Twine, Richard. "Emissions from Animal Agriculture—16.5% Is the New Minimum Figure." *Sustainability* 13, no. 11 (2021): 6276. https://doi.org/10.3390/su13116276.

United States Department of Agriculture. "Whey to Ethanol: A Biofuel Role for Dairy Cooperatives?" *Research Report 214*, February 2008. https://www.rd.usda.gov/files/RR214.pdf.

Urbańczyk, Ewelina, Maciej Sowa, and Wojciech Simka. "Urea Removal from Aqueous Solutions—A Review." *Journal of Applied Electrochemistry* 46, no. 10 (2016): 1011–29. https://doi.org/10.1007/s10800-016-0993-6.

Van der Weele, Cor, and Clemens Driessen. "Emerging Profiles for Cultured Meat: Ethics Through and as Design." *Animals (Basel)* 3, no. 3 (2013): 647–62. https://doi.org/10.3390/ani3030647.

Vaughan, Megan. "Food Production and Family Labor in Southern Malawi: The Shire Highlands and Upper Lower Shire Valley in the Early Colonial Period." *Journal of African History* 23, no. 3 (1982): 351–64.

Vaughan, Megan. *The Story of an African Famine: Gender and Famine in Twentieth-Century Malawi*. New York: Cambridge University Press, 1987.

Verbeke, Wim, Pierre Sans, and Ellen Van Loo. "Challenges and Prospects for Consumer Acceptance of Cultured Meat." *Journal of Integrative Agriculture* 14, no. 2 (2015): 285–94. https://doi.org/10.1016/s2095-3119(14)60884-4.

Vergès, Françoise. "Racial Capitalocene." In *Futures of Black Radicalism*, edited by Gaye Theresa Johnson and Alex Lubin. London: Verso, 2017.

Vespa, Jonathan. "The U.S. Joins Other Countries with Large Aging Populations." United States Census Bureau, last modified October 8, 2019. https://www.census.gov/library/stories/2018/03/graying-america.html.

Vickery, Hubert Bradford. "Liebig and the Chemistry of Proteins." In *Liebig and After Liebig: A Century of Progress in Agricultural Chemistry*, edited by Forest Ray Moulton. Washington, DC: American Association for the Advancement of Science, 1942.

Vickery, Hubert Bradford. "The Origin of the Word Protein." *Yale Journal of Biology and Medicine* 22, no. 5 (1950): 387–93.

Vickery, Hubert Bradford, and Carl L. A. Schmidt. "History of the Discovery of the Amino Acids." *Chemical Reviews* 9, no. 2 (1931): 169–318. https://doi.org/10.1021/cr60033a001.

Vince, Gaia. *Nomad Century: How Climate Migration Will Reshape Our World*. New York: Flatiron Books, 2022.

Ward, Charlotte, and David Voas. "The Emergence of Conspirituality." *Journal of Contemporary Religion* 26, no. 1 (2011): 103–21. https://doi.org/10.1080/13537903.2011.539846.

Ware, Vron. "Robbing the Soil." *Landscape Journal*, Winter 2020. https://issuu.com/landscape-institute/docs/landscape_journal_2020-1_-_12066__1_/s/10133919.

Warren, Wilson J. *Meat Makes People Powerful: A Global History of the Modern Era*. Iowa City: University of Iowa Press, 2018.

Wenger, Elizabeth. "Liver King and the Problematic Rise of Primal Manhood." *Huck Mag*, December 20, 2022. https://www.huckmag.com/perspectives/liver-king-and-the-problematic-rise-of-primal-manhood/.

Westgate, Pamela J., and Chul Park. "Evaluation of Proteins and Organic Nitrogen in Wastewater Treatment Effluents." *Environmental Science and Technology* 44, no. 14 (2010): 5352–57. https://doi.org/10.1021/es100244s.

"Whey Protein Market Size, Share and COVID-19 Impact Analysis, by Type (Isolates, Concentrates, Others), Application (Food and Beverages, Animal Feed, and Others), and Regional Forecast, 2022–2029." *Fortune Business Insights*, accessed May 30, 2023. https://www.fortunebusinessinsights.com/whey-protein-market-106555.

Wiley, Andrea S. *Cultures of Milk: The Biology and Meaning of Dairy Products in the United States and India*. Cambridge, MA: Harvard University Press, 2014.

Williams, Cicely D. "Kwashiorkor: A Nutritional Disease of Children Associated with a Maize Diet." *The Lancet* (*British Edition*) 226, no. 5855 (1935): 1151–52. https://doi.org/10.1016/S0140-6736(00)94666-X.

Williams, Cicely D. "A Nutritional Disease of Childhood Associated with a Maize Diet." *Archives of Disease in Childhood* 8, no. 48 (1933): 423–33. https://doi.org/10.1136/adc.8.48.423.

Willink, Jocko, and Leif Babin. *Extreme Ownership: How U.S. Navy Seals Lead and Win*. New York: St. Martin's, 2017.

Wolfe, Robert, Amy Cifelli, Georgia Kostas, and Il-Young Kim. "Optimizing Protein Intake in Adults: Interpretation and Application of the Recommended Dietary Allowance Compared with the Acceptable Macronutrient Distribution Range." *Advances in Nutrition* 8, no. 2 (2017): 266–75. https://doi.org/10.3945/an.116.013821.

Worboys, Michael. "The Discovery of Colonial Malnutrition Between the Wars." In *Imperial Medicine and Indigenous Societies*, edited by David Arnold. Manchester, UK: Manchester University Press, 1988. https://doi.org/10.7765/9781526123664.00013.

Worthen, Molly. "The Podcast Bros Want to Optimize Your Life." *New York Times*, August 3, 2018. https://www.nytimes.com/2018/08/03/opinion/sunday/podcast-bros-rogan-ferriss-junger.html.

Wyman, Patrick. "Bro Culture, Fitness, Chivalry, and American Identity." *Perspectives*, December 3, 2020. https://patrickwyman.substack.com/p/bro-culture-fitness-chivalry-and?utm_source=%2Fsearch%2Fbro&utm_medium=reader2.

Wynter, Andrew. *Our Social Bees: Or, Pictures of Town and Country Life, and Other Papers*. London: R. Hardwicke, 1861.

Yusoff, Kathryn. *A Billion Black Anthropocenes or None*. Minneapolis: University of Minnesota Press, 2018.

Index

Page numbers followed by *f* indicate figures.

Malo, Marta, 135

Malthus, Thomas, 114

manosphere. *See* muscular manosphere

manual labor, 41–42, 47, 58, 137–39, 139*f*

marasmus, 72, 74. *See also* malnutrition

Marcell, Taylor, 107, 108, 117

Marchesi, Greta, 41, 43, 44

Marcus, Aubrey (podcaster), 129, 130–31, 131*f*, 136, 141, 144–45

marketing: Liebig's extract of meat/beef, 46*f*, 47–48, 149; masculinity, 130–31; milk and dairy, 62–63, 67, 68, 72, 87, 101; muscle loss and, 102, 103; personal care and cosmetics, 151; plant-based foods, 2, 5, 10, 128, 141, 151, 157; protein as a multipurpose fix, 2, 3*f*, 60, 76, 129; to seniors, 121, 122*f*; whey protein powder, 82*f*, 90–91

marketing boards, dairy, 60, 87–88

Martschukat, Jürgen, 7, 110

Marx, Karl, 12, 14, 17, 43, 93

Marxism/Marxist thought, 14, 93–94

Marx's Ecology: Materialism and Nature (Foster), 93, 94

masculinity: ideal of, 129, 130–31, 134, 138–40; marketing of, 130–31, 131*f*; meat-eating and, 49, 50, 154, 157; microfascism and, 21, 129, 135–36; nutritional supplements market, 128, 129; office work and, 130, 132, 138, 143; perceived risk to, 15, 132, 136, 141, 146–47, 154; UK adolescents survey, 147; whiteness and, 15, 22, 129, 136–37. *See also* muscular manosphere

materialist philosophy, 11–17

Mbembe, Achille, 90–91, 159

McCay, Major D., 50, 51

McClousky, Maddie, 127

McLaren, Donald, 53–54, 66, 67, 72, 73, 74, 75

McNamara, Pat, 140

meat-eating: alternative proteins, 2, 9–10; demand and growth, 44–46; harms of, 1–2, 96; as ideal protein source, 39, 59–60, 156–57; masculinity and, 49, 50, 154, 157; physical labor and, 41–42; popularization of, 46*f*, 47–49, 149; racialization of, 29, 48–49, 50

meat protectionism, 10, 150, 155, 157, 158

medicalization, 102, 103–6, 110, 113, 117–18, 120, 123–24

Mendel, Lafayette, 51–52

Men Going There Own Way (MGTOW), 134

Men's Health (magazine), 131*f*

men's movement, 131–32

metabolic labor, 17, 43–44, 45, 80, 84, 92, 94–95, 99, 133, 141–42, 152

metabolic rifts and shifts, 93–95

metabolism: athletic labor and, 17, 83–84, 95, 96–97, 132–33; lack of consensus on, 4, 152; biochemical process and, 91–92, 135; quantitative analyses, 43–44; social metabolism, 17, 43, 80, 93–94; as waste disposal, 94–96, 97

methane gas emissions, 96, 97

Metzl, Jonathan, 110

Mexico, 67, 70

microfascism: emergence of, 9, 22, 129–30; patriarchal appeal of, 135–36; political continuum, 143; self-creation myth, 137, 142–43, 145–46; vision of disenfranchised masculinity, 146–47

militarization of culture, 139–40

milk, cow, 65; commerciogenic malnutrition, 72; consumption trends, 86–87; powdered-milk feeding programs, 66, 67–68, 68*f*, 69; racial purity discourse, 154

milk surpluses: cheese production, 87–88; international public health agencies, 60, 66, 67–68

misogyny, 22, 129–30, 134–35, 147

MMA (mixed martial arts) training, 140, 141

Mol, Annemarie, 18, 35

molecular biopolitics, age of, 6–8, 153

molecular science, 6–7, 26, 27–28, 32–33, 35–37, 52

Montana electoral politics, 155

Moore, Jason, 93–94, 96, 125

moralism, 7–8, 48, 74, 109–11, 112

mortality rates, 60, 66, 76, 103, 114

Moss, Michael, 87

mothers/motherhood, 56, 62–63, 64–65

Mulder, Gerardus Johannes, 30–31, 32, 33, 37

multilevel marketing (MLM), 140–41, 145

multiplicity/multiplicitous substance, 11, 17–18, 19, 21, 23–24, 29, 34–35, 80, 84, 158

Murphy, M., 115

muscle activity, 6, 32, 39–40, 45, 83

muscle loss, age-related, 15, 21, 104–7, 109. *See also* sarcopenia muscle mass, 15, 17, 21, 77, 123, 128. *See also* bodybuilding culture; fitness culture; supplement industry

muscular capitalism, 9, 158

muscular manosphere: consumerism of, 9, 140–41; defined, 129; figureheads, 129, 130–31, 144; health individualization, 145–46; microfascism of, 135, 143; militaristic ethos of, 137–40, 144; misogyny, 22, 129–30, 134–35, 147; self-care, 127, 135; self-creation fantasy, 136–37, 142, 146–47

Musk, Elon, 143, 158

Myers, Natasha, 26, 27–28, 36

MyFitnessPal (app), 5

myosin discovery, 32

mythopoets, male, 132

myths, self-creation, 136–37, 142

National Dairy Promotion and Research Board, 87

nationalism, white, 138, 139, 141

National School Lunch Act (1946), 86, 87

naturalization, 10, 16, 19, 109, 117, 118

Nature, 74, 107

Nature-Society dualism, 84–85, 93–94

Navy Seals, US, 137–38

neoliberalism: in decline, 129–30, 135; financialization and, 115, 117–18, 124; fitness culture, 7–8, 11, 22; precarious labor, 145; healthcare spending cuts and, 125; healthism, 109–13, 121, 149, 153; life sciences research, 120; physical development, 55. *See also* muscular manosphere

Nestlé (company), 62–63, 72, 121

Nestlé Health Science, 119, 121

networked culture, 134–35

neurochemical selfhood, 7

New Age Eats, 10

New Mexico symposium, 104–5

New York Times, 87, 130

NHS (National Health Service) (England), 116

nitrogen content, 3, 4, 35, 38, 45–46, 52. *See also* meat-eating

nitrogen cycle, global, 15–16, 97–98

nitrogen fertilizers, 44, 97–98

nitrogen pollution, 15–16, 81, 89, 96–98, 99

Nomad Century: How to Survive the Climate Upheaval, 124

nomenclature, 30–31, 32–33

Nott, John, 54, 58, 61, 62, 63, 64–65, 76; on protein gap, 72; on waste techno-supplements, 71

nursing mothers. *See* breastfeeding

nutricentric subjects, 7, 21

nutrients, 4–7, 21, 25, 37–40, 49, 52

nutritional epigenetics, 94

nutritional science: albuminates in, 41, 42–43; careerism, 74, 75; dietary guidelines, 4–5; lack of consensus, 152; *protein* coined in, 30–31; protein folding and amino acids, 35–37; protein intake debates, 68, 151–53; protein primacy, 19–20, 56; sarcopenia, 102; whey protein value in, 20–21, 81, 90. *See also* malnutrition

nutrition-industrial establishment, 74, 75

Nutrition Insight, 103, 121

Nutrition in the British Colonial Empire (1939), 63–64

nutritionism: colonial medicine, 60–62; described, 6–7, 8, 29; global inequality and, 6, 20, 56–57, 72–74, 111, 113; protein intake debates, 4–5, 152–53; protein primacy and, 41–43, 44. *See also* Liebig, Justus von

Obama, Michelle, 87

office as social space, 133, 138

omega-3, 55

Omnit (brand), 130, 131, 141

On Garbage (Scanlon), 84

Ontario and whey dumping, 89

ontology, protein, 25*f*, 37, 150. *See also* proteinous matter

optimization discourse, self-, 22, 111–12, 124, 128, 129, 130–31, 133, 142, 144, 146. *See also* muscular manosphere; nutritional supplements

Oregon livestock exposition, 48*f*

organic chemistry, 34, 37, 38–40, 43, 150. *See also* Liebig, Justus von

Orientalism, 51

Ormsby-Gore, William, 59

Orr, Boyd John, 59, 60

osteoporosis/osteopenia, 105

ovalbumin, 28

Overend, Alissa, 153

quantitative analyses, 41, 42–44, 52

racialization, 58, 61, 114, 118
racial superiority discourse, 20, 29, 48–49, 50
racism, 62, 65, 75, 136–37, 154
Raimondo, Gina, 115
Rajan, Sunder, 120
recommended daily allowance (RDA), 4–5, 151–52
recommended dietary intake (RDI), 119
reductionism, 6, 29, 43, 64–65, 153, 160
relationality, 14, 15–16, 19, 51, 93–95, 145, 146
Remski, Matthew, 140–41, 144, 145
Rendering Life Molecular (Myers), 26, 27–28, 36
Republicanism, 155, 157
responsibilization discourses, 20, 77, 102, 110, 124, 129, 138, 145–46
restoration myth, 135–36, 137, 142
Reynolds, Jacqueline, 32
right-wing movements, 135–36, 147, 154, 155, 157–58
risk discourse, 112, 117–18, 149
Rockefeller Foundation, 69
Rogan, Joe, 129, 130, 138, 140, 141, 143, 144, 145, 154
Rose, Nikolas, 6, 91
Rosenberg, Gabriel, 157
Rosenberg, Irwin, 104–5, 106–7, 117, 120
Roy, Parama, 50–51
Ruanda-Urundi, 67

Sai, Fred, 72
Santos, Federico Gómez, 70
Sarconeos (BIO101) (drug), 122
sarcopenia: clinical guidelines, 108–9; defini-tions, 21, 102, 105–8; disease classification, 77, 102–3, 104–6, 113, 116–17; disease etiol-ogy debates, 107–9, 123; drug development, 120–21, 122; economization of, 117–18, 119–20; reductionism, 125; risk discourse, 112; university grant funding, 120–22
Sarmiento, Eric, 19
Sathyamala, C., 72
Scanlan, John, 84
Schmidt, Carl, 34
school lunch programs, 5, 86, 87
Schwarzenegger, Arnold, 108

science. *See* nutritional science; protein sci-ence; *individual figures*; *specific fields*
Scott-Smith, Tom, 54–55, 68, 71, 74, 76
Sears, J. P., 141, 144
self-actualization. *See* optimization dis-course, self-
self-creation myth. *See* autogenetic sovereignty
selfhood, molecular, 6–7, 8
selfhood, responsible, 111–112, 129–31, 140–41, 144, 146–47
Semba, Richard, 53, 75
senescence, 105–6, 107, 109, 117, 132
Sentmanat, Tony, 140
sewage, 71, 89
sexist tropes, 154, 155
Sexual Politics of Meat (Adams), 96
Silicon Valley, 10, 154, 158
Singapore, 63
Sliced Bread (BBC radio series), 159
Smithers, Geoffrey, 79, 80, 86
Snowpiercer (film), 155
social determinants of health, 71, 109
social gerontology, 109
social infrastructure, 123–24, 145, 157–58
social media, 9; digital platforms, 128, 133, 134–35, 147; influencers, 128–29, 130–31, 137–38, 154. *See also* podcasters
social metabolism (*Stoffwechsel*), 17, 43, 80, 93
social movements: environmentalism, 90, 93, 155, 159–60; global food insecurity and, 72–75; men's/father's rights, 131–32, 134; networked cultures, 135; vegetarianism, 50–51, 76
social relations, 12, 136–37
socio-ecological relations, 15–17, 91, 98–99, 142, 159
sociology of medicine, 106
soil chemistry, 43, 89, 93, 97
sovereign self-making, 9, 128, 136, 137, 142–46
"soy boy" epithet, 154
SpaceX (Musk), 158
sports nutrition market, 128, 132, 137–38
standards and standardization, 4–5, 6, 49–50, 51
Stannus, H. S., 60–61
stratification, social, 47, 109, 118, 155
structural adjustment policies, 76, 102

whey: historical use of, 79–80; refining process, 15–16, 83, 98, 142; reuse efforts, 88, 89, 89*f*; toxicity of, 15–16, 21, 81, 85, 89, 96–98, 99; (bio)value, 20–21, 80, 81, 90, 99, 151; as waste, 81, 85, 99. *See also* protein powder, whey

whey dumping, 89–90, 98

whiteness, 22, 129–30, 136–37

white supremacy/nationalism, 134, 136–38, 141, 143, 154

who (World Health Organization). *See* World Health Organization (who)

Williams, Cicely, 60–64, 71

Willink, Jocko, 137–38, 139*f*, 140, 141, 144–45

Wisconsin whey dumping, 89*f*

Wöhler, Friedrich, 31, 32

women: aging population, 116, 118, 119; infant feeding politics, 42, 62–63; population growth intervention, 114; precarious labor, 145;

woodsman, study of German, 42

Worboys, Michael, 58

work-space casualization, 133

World Bank, 70

World Food Conference (1974), 73*f*

World Health Organization (who): disease classification by, 102, 106, 120–21; icd-mc Diagnosis Code, 117, 120–21; kwashiorkor, 66; protein intake guidelines, 4, 75; who/fao Expert Committee on Nutrition, 65–67, 68–69, 70–71, 74. *See also* pag (Protein Advisory Group)

world-historical processes, 11, 12–14

world wars and milk consumption, 86

Wyman, Patrick, 138–140

Wynter, Andrew, 45, 48, 53

Yale University research, 50–51

YouTube, 128, 133, 138, 140, 141, 143

Zimmerman, John, 42

9 781478 032922